AF749495

Haim Goren is Professor of Historical Geography, Tel-Hai College, Israel. He has a longstanding interest in the Holy Land, European activity in Ottoman Palestine and the Near East, and the history of the modern scientific study of these regions. His books include *Dead Sea Level: Science, Exploration and Imperial Interests in the Near East* (2011) and (with E. Dolev and Y. Sheffy) *Palestine and World War I: New Perspectives* (2014), both published by I.B.Tauris.

Jutta Faehndrich is a cultural historian and researcher with the Leibniz Institute for Regional Geography, where her research focus is on the history of cartography, European cultural history, and cultures of memory.

Bruno Schelhaas is Head of the Archive for Geography, Leibniz Institute for Regional Geography, Leipzig, Germany. His interests include the history of geography and cartography, historical geography and archival science.

'*Mapping the Holy Land* is a scholarly yet accessible work that presents for the first time a detailed account of the activities that gave rise to the first scientific maps of Palestine, in the nineteenth century. The authors examine the work of three major figures – Augustus Petermann, Charles Meredith van de Velde, and Edward Robinson – and the influence that German geographical thought of the period had upon them. Drawing upon recently discovered archival material, the authors have produced a work that will enrich and engage all those with an interest in the development of maps and the Holy Land.' – **Gideon Biger, Professor Emeritus, Department of Geography and Human Environment, Tel Aviv University**

'This book by three authors who are experts in their fields of archival science, the history of cartography, historical geography, and cultural history, sets an exemplary standard for detailed research into archives which have been hitherto unexploited for their content as far as the mapping of the Holy Land in the nineteenth century is concerned. Much is revealed, not only of the map compilation methods and commercial map publishing practices of those times, but we also learn of the more elusive human stories behind what were the ground-breaking cartographic products of their time for this area. There is a cornucopia of new material here, which will be relevant to a readership that goes beyond the geographical limits of the Holy Land. This presentation of substantial original research into the major nineteenth-century German map publishers and mapmakers, and their associates, provides an inestimable service to all students of map history whatever their geographical focus. This book must rank as a major contribution to the subject of the history of cartography in general, as well as an essential reference for the mapping of the Holy Land in the nineteenth century in particular.' – **Dr Yolande Hodson, Formerly Honorary Secretary of the Palestine Exploration Fund**

TAURIS HISTORICAL GEOGRAPHY SERIES
Series Editor: Robert Mayhew, University of Bristol

Published and forthcoming in the Series:

1. *Zambesi: David Livingstone and Expeditionary Science in Africa* by Lawrence Dritsas
2. *New Spaces of Exploration: Geographies of Discovery in the Twentieth Century* by Simon Naylor and James R. Ryan (eds)
3. *Scriptural Geography: Portraying the Holy Land* by Edwin James Aiken
4. *Bringing Geography to Book: Ellen Semple and the Reception of Geographical Knowledge* by Innes M. Keighren
5. *Enlightenment, Modernity and Science: Geographies of Scientific Culture and Improvement in Georgian England* by Paul A. Elliot
6. *Dead Sea Level: Science, Exploration and Imperial Interests in the Near East* by Haim Goren
7. *Nature Displaced, Nature Displayed: Order and Beauty in Botanical Gardens* by Nuala C. Johnson
8. *Geography and the Classical World: Unearthing Historical Geography's Forgotten Past* by William A. Koelsch
9. *Imagining the Arctic: Heroism, Spectacle and Polar Exploration* by Huw Lewis-Jones
10. *The First Mapping of America: The General Survey of British North America* by Alex Johnson
11. *Mapping the Holy Land: The Foundation of a Scientific Cartography of Palestine* by Haim Goren, Jutta Faehndrich and Bruno Schelhaas
12. *The Idea of Central Europe: Geopolitics, Culture and Regional Identity* by Otilia Dhand

MAPPING THE HOLY LAND

The Foundation of a Scientific Cartography of Palestine

HAIM GOREN, JUTTA FAEHNDRICH
AND BRUNO SCHELHAAS
with
PETRA WEIGEL

I.B.TAURIS
LONDON · NEW YORK

Published in 2017 by
I.B.Tauris & Co. Ltd
London • New York
www.ibtauris.com

ISBN: 978 1 78453 454 7
eISBN: 978 0 85772 983 5
ePDF: 978 0 85772 785 5

Tauris Historical Geography 11

A full CIP record for this book is available from the British Library
A full CIP record is available from the Library of Congress

Library of Congress Catalog Card Number: available

Typeset by Saxon Graphics Ltd, Derby

Contents

Figures, Plates and Tables

FIGURES

PLATES

TABLE

Preface and Acknowledgements

This book is a result of the German–Israeli research project 'Robinson, van de Velde, and German Holy Land Cartography in the Mid-Nineteenth Century', funded by the German–Israeli Foundation for Scientific Research and Development (GIF) and realised at Tel-Hai College (Upper Galilee, Israel) and the Leibniz Institute for Regional Geography (Leipzig, Germany). The Principal Investigators (PI) of the project, Haim Goren (Tel-Hai College) and Bruno Schelhaas (Leibniz Institute for Regional Geography), together with research assistant Jutta Faehndrich (Leibniz Institute for Regional Geography), are the authors of this volume. Petra Weigel (Gotha Research Library) is responsible for the selection of most of the images (maps and archival documents of the Perthes Collection). The book is divided into three main chapters, presenting three case studies of our topic: the reconstruction of the history of German Palestine cartography in its decisive period between 1830 and 1875.

Both institutions at Tel-Hai and Leipzig supported the project generously. The Tel-Hai MIGAL Research Fund, together with the college management led by former President Yona Chen, enabled Haim Goren to make a number of visits to Germany, which were of crucial importance for the project. It also hosted a workshop, visited by everybody connected with it. The Leibniz Institute for Regional Geography, especially the working group on History of Geography as well as the Central Geographical Library and the administration supported the project from the very beginning in all organisational and scientific issues with personal engagement and expertise. Finally, the Fritz Thyssen Foundation in 2012 approved an additional four months Herzog-Ernst-Scholarship to Jutta Faehndrich for realising intensive archival work at Gotha Research Library.

During the more than four years spent on the project (2011–15), the research team benefited from a great number of colleagues and friends in Israel, Germany and elsewhere. We are very thankful to all of them. Firstly, we have to thank our student assistants Sophie Perthus and Teresa Kilian, who contributed basic work packages, in particular with very detailed source collection, map analysis, deciphering and transcribing of handwritten documents. The Cooperating Investigators Heinz Peter Brogiato and Ute Wardenga (Leibniz Institute for Regional Geography) and Rehav (Buni) Rubin (Hebrew University of Jerusalem) joined the project from the very beginning and promoted it with their expertise and long-standing experience in international research at all times.

In Israel, we found essential support from several experts. In the first place we have to thank research assistant Moshe Maggid (Jerusalem) and student assistant Ana-El Badach (Jerusalem). Shalom Tarmachi (Tel-Hai College, Map Collection), Senior Cartographer Tamar Soffer (The Hebrew University of Jerusalem) and Ayelet Rubin (Eran Laor Collection, The National Library of Israel, Jerusalem) opened their collections for us and gave precious assistance, too.

We managed to find and analyse a unique pool of published and especially archival sources in primarily German, Israeli, British and US collections. The findings exceeded all our expectations. Accordingly, the cooperation with the collections was the basis for the success of our project. The Perthes Collection at Gotha Research Library has been our most important partner for many years. The realisation of our project was only possible with the help of the unique and rich documents of the collection. Petra Weigel, head of the Perthes Collection, was an active member of the project team from the beginning. She supported our research, participated in our workshops and conferences and she was active in writing papers on the subject. Her dedication and expertise greatly advanced our project. She and the collection's archivist Sven Ballenthin supported our research, answered any kind of requests and opened the collections generously and with patience. We owe a tremendous debt of gratitude to both, to the whole staff of Gotha Research Library and especially to its director Kathrin Paasch.

The second collection in Germany with highly important sources for our subject is the Berlin State Library. We are especially grateful to Wolfgang Crom and Markus Heinz (Map Department) and to Jutta Weber (Manuscript Department).

Likewise, we have to thank friends and colleagues in the USA, in particular Christian Goodwillie, director and curator of Special Collections and Archives at Hamilton College Library (Clinton, NY) and his team, Jay G. and Hermine Williams, who managed the Edward Robinson Papers for many years, and the staff of Houghton Library, Harvard University (Cambridge, MA).

We are also grateful to Jakob Eisler (Landeskirchliches Archiv Stuttgart), Bruno Tauché (University Halle-Wittenberg), Christopher Fleet and Karla Baker (National Library of Scotland, Edinburgh) and to many, many friends for supporting us so warmly. Finally, we want to thank I.B.Tauris Publishers and in particular Senior Editor David Stonestreet and his team, and copy-editor Christopher Shaw, for realising this book.

Haim Goren, Jutta Faehndrich and Bruno Schelhaas

Introduction: Mapping the Holy Land

Haim Goren, Jutta Faehndrich and Bruno Schelhaas

Much of the research dealing with nineteenth-century Palestine describes the period as the 'century of change'. The territory underwent a process of transformation from an underdeveloped, sparsely populated and neglected province of the Ottoman Empire to a country on its way to modernisation. Along with local Ottoman reform processes, one of the factors in this transformation was the influence of Europeans, which took many forms. This included religious, diplomatic and economic activity, European immigration and settlement, and intense field research of a country that had hitherto not been the focus of European scientific exploration.

The first decidedly modern scientific investigations of the Holy Land began in the middle of the eighteenth century but the real increase in European activity happened after the 1830s. The early stage was followed mainly by sporadic and intermittent studies of a general nature. Limited in scale and scope, these works were initiated and conducted by individuals or small groups, including religious and social organisations, and only occasionally supported, though not initiated, by governments. Scholars from Europe – mainly Great Britain, France and Germany, Russia and North America – were the main protagonists. As the century advanced, discovery and exploration gave way to programmed scientific research, which eventually focused on special topics or regions.

The history of Holy Land cartography followed the broad contours of this process. Earlier 'modern' maps suffered many deficiencies, mainly due to a reliance on partial data that were neither precise nor systematic. However, over the course of the nineteenth century, an increasing body of material, based on relatively reliable and precise measurements, was collected and the maps accordingly became more accurate.

The mapping of the Holy Land was greatly shaped by German scholarship, either by German scholars themselves or by others who were strongly affiliated to, and influenced by, the German cultural and scientific community. The two outstanding figures within this were the American Biblical scholar and minister Edward Robinson and the Dutch naval officer, cartographer and landscape painter Charles William Meredith van de Velde, both active from the late 1830s to the 1860s. Both undertook their own survey, made their own measurements and noted down their itineraries. In combination with already existing knowledge, their data thus generated was used to produce new maps of the region. The German cartographer and science manager August Petermann is the third central protagonist described in this volume. He participated in the British debate on the Dead Sea level and the features of Palestine's physical geography in the 1850s, and he produced a number of influential maps of the region. Moreover, there are interesting connections between the three individuals involved.

In the case of Edward Robinson, who cooperated closely with his companion Eli Smith, we were able to reconstruct a highly complex network of agents and an often complicated collaboration with the cartographer Heinrich Kiepert when it came to producing the maps. The cooperation was international, but with a strong German influence, very closely connected to Heinrich Berghaus, Carl Ritter and the German scholarly community in Berlin, Leipzig, Halle and Göttingen.

The situation was very favourable for a historical study: besides a great number of basic printed sources – in particular Robinson's travel books and research reports, Kiepert's map memoir and Kiepert's main maps – we were able to identify and make use of a unique body of archival sources, many of them representing the very first time by any scholar. The Edward Robinson Papers at Hamilton College Library (Clinton, NY) and the Eli Smith Papers at Houghton Library, Harvard University (Cambridge, MA) were certainly most important for this task. It was both a challenge and a unique opportunity to have access to more than 1,250 relevant documents, mainly handwritten, including an intense correspondence between Robinson and Smith during the writing of their book and production of the maps. Both had almost unreadable handwriting, which called for careful and professional deciphering (the use of '[?]' in some extracts shows where it was completely unreadable). With regard to the German collections it was, in particular, the Berlin State Library that offered an extensive range of relevant documents concerning our topic. The Map Department holds several Palestine sketch maps and published maps created by Heinrich Kiepert, and the papers of Carl Ritter and Heinrich Berghaus, which offered much contextual information, can be found in the Manuscript Department. Much of this archival material is referred to and quoted in Chapter 1.

In contrast to Edward Robinson and his network, C. W. M. van de Velde was a highly independent scholar, researcher and cartographer, combining many skills in one person. His fieldwork was planned strategically to get all the necessary data and material for his *Map of the Holy Land*. Again, his knowledge transfer and map-making were international, but carefully chosen. The cooperation with Justus Perthes' publishing house in Gotha originated in van de Velde's decision to produce his *Map of the Holy Land* at the best possible quality.

The documents found in the unique Perthes Collection at Gotha Research Library provided the basis for studying the making of this map in a most detailed way. After a close reading, decipherment and transcription of several hundred pages of correspondence between van de Velde and his publishers, including third parties, we were able to reconstruct the history of knowledge production with an emphasis on the intense exchange between van de Velde and Justus Perthes. The holdings of the Perthes Collection yielded a number of new facts regarding the map-making process. Surprisingly, August Petermann was not involved in making the 1858 *Map of the Holy Land.* Instead, the managing director Adolf Müller, one of two managers who took over after the untimely death of senior director Bernhardt Perthes, discussed and decided all details with van de Velde.

Chapter 3, on August Petermann's Palestine maps, shows a highly talented cartographer and science manager, strongly influenced by the German tradition, in direct contact with Alexander von Humboldt and Heinrich Berghaus, but also influenced by British development. He never set foot in the Holy Land but he was able to collect and combine all available information, often exclusively, as the basis for several new Palestine maps. Petermann's contributions to Palestine cartography date back to his Edinburgh years in the mid-1840s and became more important after his move to London in 1847. Here he established his own cartography business and produced a series of atlases in very short time, all of them containing Palestine maps. In Petermann's years in Gotha at Justus Perthes, from 1854 until his death in 1878, Palestine was not so much the focus of his work. Nevertheless, the few Palestine maps in his journal, the *Mittheilungen*, and in particular the sheets for *Stielers Hand-Atlas*, attest to Petermann's unique method of operation, including an interpolation of all available knowledge and a strategic crossover marketing of his maps. Again, the Perthes Collection was the most important source for our analysis. Most of the documents were available for scholarly use for the very first time.

In the mid-nineteenth century, cartography and geography, as well as the emerging Biblical Geography were in the middle of a professionalisation process. The competitive relation between the various European agencies and protagonists involved in making and publishing maps of Palestine is a central theme within the following chapters. Obviously, the most important ideology and belief behind mapping Palestine was its character and quality as the Holy Land. On the one hand, there is Charles William Meredith van de Velde, a surveyor and cartographer trained with the Dutch navy who decided to measure Palestine single-handedly, driven by Protestant zeal. He was a generalist and amateur Biblical Geographer, who did not, however, meet the later criteria of an academic scholar. On the other hand, there is the American Edward Robinson, founding father of Biblical Geography as a scholarly discipline, who published the maps to accompany his Palestine research in Germany with Heinrich Kiepert, who would become the foremost German cartographer of the nineteenth century. Finally, there is August Petermann, the highly talented cartographer and organiser, who was able to collect all the available and the latest information about Palestine research, often exclusively. *Mapping the Holy Land* presents three different approaches to Biblical Geography, each with a different strategy of visualisation and knowledge transfer.

1 Edward Robinson, Eli Smith and the Cartography of Berghaus and Kiepert

Haim Goren

I have not time to relate where we have been & what we have seen. But I [...] shall be able to throw considerable new light [...] on the wonders of God.

Eli Smith to Brother, Alexandria, 12 July 1838, ESP, ABC 60 (105)

INTRODUCTION

The expeditions and publications of the American biblical scholar Edward Robinson and the Presbyterian missionary Eli Smith mark the beginning of the modern scientific research and survey of Palestine, the 'Holy Land'.[1] The elucidation and visualisation of the Holy Bible were Robinson's highest ambition. His close relations with the German cultural, scientific and social world brought the team's scientific endeavours to the international arena. Robinson adopted many ideas and theses of the leading German geographer Carl Ritter, who he met while studying at the universities of Halle and Berlin. He worked with the highly influential German cartographers Heinrich Berghaus and Heinrich Kiepert, and the latter became the cartographer for all of Robinson and Smith's works.

This chapter focuses on the cartographical aspects of Robinson and Smith's journeys and publications. It is based on previously published material, very often contradicting the findings reported,[2] and on newly discovered archival sources found in Robinson's papers held in the Special Collections at Burke Library, Hamilton College (Clinton, NY), and in Smith's papers at Houghton Library, Harvard University (Cambridge, MA). Additional written and archival sources are located in Germany, primarily in Gotha and Berlin, and in London.

The background to Berghaus's and Kiepert's cartography, including Robinson's and Smith's personal histories and education, their pre-travel preparations, travel routes and post-travel publications, and their reception, will be presented in brief. A forthcoming research project, titled 'Robinson and Smith, Generators of Change in Holy Land Scientific Study: New Considerations' will concentrate on the aforementioned topics. Here I touch only upon those that are of importance to the production and publication of the maps.

This chapter addresses the following issues and research questions:

- Ritter's role in the genesis of ideas, scientific development and the work of Robinson;
- Berghaus's work with Robinson;
- when and how Kiepert entered the scene replacing Berghaus;
- the decision making process that led to the collaboration between Kiepert and Robinson, and when, why and how it began;
- their exact work-plan and methods;
- Robinson's contribution to Kiepert's technical work; and
- the exact nature of Ritter's role in the publications and maps.

In the following pages, I have attempted to put these issues concerning the involved scholars and experts, processes and developments, and mutual studies and works, as well as the quarrels and rivalries, into their place in the narrative, whose primary participants were Robinson, Smith, Ritter, Berghaus and Kiepert.

ROBINSON AND SMITH: THE MAKING OF A SCIENTIFIC PARTNERSHIP AND THE FIRST EXPEDITION

After finishing his studies in the first class of Hamilton College, which was established in 1812, Edward Robinson enrolled at Andover Theological Seminary, a stronghold of Protestant conservatism (see Fig. 1.1).[3] Since the founding of the school in 1809, Hebrew instruction had been a cornerstone of its text-oriented philological approach to Bible study. Robinson, a student and protégée of Moses Stuart,[4] Andover's first and leading Hebraist, was appointed instructor of Hebrew after studying the language for only one year. It was Stuart's idea to send Robinson for four years of advanced study, especially Hebrew study, in Europe, where he spent most of his time at the German universities of Halle and Berlin. Upon returning to Andover Theological Seminary, he wrote:

> My duties will consist in lecturing in the department of Sacred Literature in connexion with Prof. Stuart; in having charge of the library; & than in having the superintendence of the so-called *Cadman Press* belonging to the Seminary. Among other things I am to be the Editor of a quarterly journal consisting partly of original & partly of selected activity, mostly relative to biblical Literature, though not exclusively so.[5]

Robinson served as 'Professor Extraordinary of Sacred Literature and Librarian, in Andover' from 1830 until 1833. He left Andover to research independently in Boston and New York and wrote: 'the three years that I have resided in this city, have been spent in literary labours having reference to Biblical Studies, the preparation of a Greek Lexicon of the New Testament, & a translation of Gesenius Heb. Lexicon.'[6] In December 1836, he 'received from the Directors of the New York [Union] Theological Seminary an anonymous invitation to take charge of the Professorship of Biblical Literature in that institution with a salary of $2500 per annum'. This appointment at the UTS led to his first visit to the Holy Land in 1838.[7]

In contrast to the majority of historians, who wrote about Robinson without consulting primary sources,[8] Jay Gomer Williams and his wife, Hermine, had Robinson's papers in

Figure 1.1 Daniel Huntington, Edward Robinson, D. D., L. L. (1794–1863). Class of 1816, 1858. Oil on canvas, 30 × 25 in. (76.2 × 63.5 cm)
Source: Ruth and Elmer Wellin Museum of Art at Hamilton College, Clinton, NY.

their possession for twenty years, relying upon them for their books, before transferring them to their current location. Williams read about and analysed thoroughly the period of Robinson's studies in Germany and elsewhere from '400 pages of letters written between 1826 and 1830'.[9] Williams focused on Robinson's development as a theologian and its influence on his scientific activity and publications, which included many philological works, some of them of significant value.[10] Hermine Weigel Williams transcribed and edited Robinson's letters to his sister, Elisabeth, written between June 1826, when he sailed from New York, and May 1829, when Robinson and his newly married German wife, Theresa Albertina Louise von Jakob (TALVJ), returned to Halle after a 41-week honeymoon in Europe. Hermine wrote a short biography of Robinson, adding hers to the 'classical' biography published by Smith and Hitchcock in 1863 and to the longer and more detailed one in her husband's book.[11]

In the letters, as well as in Jay Williams' books about Robinson's studies in Germany during the second half of the 1820s, we encounter the names of many German scholars with whom he developed friendly as well as scientific connections. Most frequently mentioned are the orientalist and Hebrew lexicographer Wilhelm Gesenius from Halle;[12] the theologian of Jewish origin Johann August Wilhelm Neander from Berlin; theologian August Tholuck from Halle; Emil Rödiger, expert in oriental languages and co-worker of Gesenius and Moses Stuart; Ernst Wilhelm Hengstenberg, neo-Lutheran theologian and expert in Arabic; and publishers such as Karl Tauchnitz from Leipzig. This is only a partial list of Robinson's academic, professional and social network in Germany; some of these scholars were also his teachers. Admittedly, he was helped by the social position of Theresa and her family, led by her father, the political economist Ludwig Heinrich von Jakob, but there is much more to the story. These leading scholars found in Robinson a competent and most interesting and innovative scientist who contributed much to their knowledge and research. Of them, the geographer Carl Ritter from Berlin (as discussed later in the chapter) turned out to be the most important figure for Robinson's journeys and publications.

From his first stay in Germany, Edward Robinson saw it as his mission to purchase and collect as many 'Palestine' books as possible, primarily of history, geography and travel journals, for Andover's Divinity School Library. Leipzig, with its many publishers and booksellers and the leading book fair, turned out to be ideal for this purpose.[13] When he was appointed librarian at UTS, book purchasing took an increasing amount of his time, also when he travelled to Germany on his way to and from the Holy Land in 1838. Various letters in his archive reveal a long negotiation, from as early as 1837, for the purchase of the Rosenmüller library for UTS. The Rosenmüllers, both father and son, Johann Georg and Ernst Friedrich Karl, were orientalists and theologians who lived and taught in Leipzig; the latter was the editor of the catalogue for Ulrich Jasper Seetzen's oriental manuscripts.[14] Robinson tried to acquire the library through the mediation of the Leipzig publishing house J.A. Barth, named after its founder, Johann Ambrosius Barth, and managed by his son, Wilhelm Ambrosius. According to the last letter dealing with the desired purchase, dated 6 March 1840, just before he left Berlin, Robinson did not succeed.[15] But upon returning to New York from Germany later that year, he 'brought home three or four thousand more [books]'.[16] These were a significant addition to the existing UTS library holding 14,000 volumes, but it seems that Robinson held most of them in his house, as after his death many reached the Burke Library at Hamilton College.[17] Williams also describes Robinson's later acquisition of 'several thousand books' from the collection of the Catholic theologian Leander (Johann Heinrich) van Ess, whom he visited in Darmstadt in September 1837.[18]

Robinson found a worthy research and expedition partner in the Presbyterian Reverend Eli Smith. Smith graduated from Yale in 1821 and taught for several years in Georgia. He then enrolled at Andover Theological Seminary, where Robinson was his teacher. He graduated in 1826.[19] After graduation, he joined the American Board of Commissioners for Foreign Missions (ABCFM; established in 1810)[20] and went to Malta and then to Syria (1827) in order to improve his Arabic skills and to promote the translation of the bible into Arabic. In 1828, due to the hostilities of the Greco-Turkish War, Smith was forced with other missionaries to leave Beirut. He travelled to Greece and, in 1831, he joined his Andover colleague Harrison Gray Otis Dwight on a missionary research excursion in Asia Minor.[21] In 1832, after returning from Armenia, Smith paid a short visit to Andover where he met with Robinson, his former teacher. They had already discussed the possibility of a trip to the Holy Land. Smith expressed his utmost desire 'to be your travelling companion' in Palestine and did not hide his eagerness to accompany Robinson, in spite of his obligations to the missionary society.[22] It took them six more years to realise this desire.

In 1833, Smith resettled in Beirut to carry on with his missionary duties. His home in Beirut became a focal point for Western travellers to the Near East.[23] His profound knowledge of, and acquaintance with, the Levant, its customs and languages and, perhaps most important, his relations with local bureaucrats, foreign diplomats and missionaries in the Middle East were undeniably already useful on his 1834 solo excursion to the Holy Land.[24] Sarah L. Huntington Smith, Eli Smith's wife, noted this advantage in her memoirs. The 1834 voyage was originally undertaken for missionary reasons, but became important for Smith's research. While abroad, Smith prepared his list of 'Names of Places, chiefly in the Pashalik of Damascus, obtained during a Journey in the Spring of 1834'.[25] Sarah Smith died in 1836, before her husband's expedition with Robinson.

Smith turned out to be the ideal 'man in the field'. He was exceedingly familiar with the Levant, had a vast knowledge of Oriental languages and experience in scientific research.[26] A remarkable example of the contribution of their close scientific partnership is found in Robinson's letter to the British naval officer and geographer John Washington, where he wrote, 'my companion was the Rev. Eli Smith, American missionary at Beirut, who is probably known [...] he has been 12 years in the Levant, chiefly in Lebanon, and speaks the *area languages*.' He admired Smith, calling him 'my fellow teacher in Palestine'.[27]

Having as his companion this well-known biblical and linguistic scholar who was at the same time deeply rooted in the Levantine mentality allowed Robinson to execute his research without the assistance of the locals, as his contemporaries were forced to do. Indeed, by 1834, Smith had been acknowledged as an expert in local affairs, as demonstrated by his correspondence with John William Perry Farren, British Consul-General to Syria, whose seat was to be in Damascus.[28] Farren was nominated at the end of 1830 and arrived in Beirut six months later. Strong opposition from the local population in Damascus, followed by threats to assassinate him, forced Farren to remain in Beirut. He settled in Damascus only in January 1834, following its conquest by the Egyptians.[29]

Robinson and Smith's joint expedition to the Holy Land not only yielded fertile results in the areas of bible and geography, it advanced the field of linguistics, which was Smith's '*Schwerpunkt*'. Toponomy, the connection between contemporary and Biblical-historical place names and their correct pronunciation, had been a major research area in the study of the Holy Land and, of course, in reconstructing its Scriptural Geography, an issue worthy of its own detailed study. Robinson often acknowledged Smith's expertise in the field, as in the opening to the *Biblical Researches*:

> He had in former days been my pupil and friend; and a visit to the Holy Land naturally became a topic of conversation between us [...] I account myself fortunate in having been thus early assured of the company of one, who, by his familiar and accurate knowledge of the Arabic language [...] indeed, to these qualifications of my companion, combined with his taste for geographical and historical researches, and his tact in eliciting and sifting the information to be obtained from an Arab population, are mainly to be ascribed the more important and interesting results of our journey.[30]

Smith's research methods and rules of orthography were adopted by the ABCFM, as well as by Robinson in his books and maps.[31] With regard to Smith's rules of orthography, Robinson's biographers, his UTS colleagues Henry Boynton Smith and Roswell Dwight Hitchcock, quoted him as saying 'since Ritter has pronounced its encomium, its authority is sealed, and its fame is fixed'.[32]

With Robinson in New York and Smith in Beirut, naturally they had to prepare for the expedition by letter, a slow, tiresome and unsecure procedure, but they did not seem to have any other choice. They arranged for their equipment and financing. In one of his letters, Robinson, at that point in Egypt and 'unable to find a compass', asked Smith to bring along his instrument and try to purchase another one, 'either at Smyrna or anywhere else'.[33] In another letter from Cairo, he listed various technical issues as well as additional equipment needed for the expedition, including a firearm (government requirement) for travelling through the more 'wild' areas such as the Dead Sea, wooden boxes, a proper servant, a guide for Sinai and information concerning the route to Petra from Gotthilf Heinrich von Schubert and Lord Alexander William Crawford Lindsay, who had visited the legendary Nabatean city in 1836.[34] Smith purchased the necessary instruments and gear. They included 'a piece of painted canvass to defend your bed from the rain & from the dampness of the ground', a pocket compass, circumferentor (used to measure horizontal angles, preceded the theodolite on survey), a compass, measuring tape of 60 ft, thermometer, sextant, etc. However, 'an artificial horizon & chronometer are not to be found'.[35]

In a letter to Smith dated January 1839, Robinson first mentioned technical difficulties with their measuring instruments and the problems stemming from their lack of expertise in using them. Robinson, already busy writing the book, related his difficulty in understanding the bearings they took, which 'lies, I apprehend in the unperfection of the instrument as to the measurement of any *long* line, & small angles'. He was concerned about 'the angles of elevation which we took, [...], especially as the results did not accord well with our estimates'.[36]

Their exceptional methods of travel and inquiry have been described in prior studies. Ritter was one of the first scholars to elaborate upon the reasons behind the precision, uniqueness and authenticity of their study:

> Following their uniform plan of travels, Robinson and Smith did not lodge in the convents, but in the open air, or in the houses of the people, employed the Syrians as their guides, and struck across the country through the most retired and unexplored byways. Nor did they ask direct questions, which usually get the answer which the Arab thinks the questioner wants; but by the most indirect interrogatories and cross questions, and by comparing the answers gained from different persons, they at least felt, in most cases at least, that they had in some measures attained the actual facts. The services of MR Smith, who had for many years been a missionary in Syria, and was perfectly familiar with the popular speech, were indispensable.[37]

Upon finishing their travels in the Holy Land, they went together to Europe. Robinson visited his family in Halle and then continued to Berlin to write the book. Smith went to nearby Leipzig to prepare his letters for the Arabic typography. Robinson wrote: 'The specimen of Arabic type is beautiful. It is unlike any other yet cast, & will, I doubt not, make quite a sensation in the literary Arabic world.'[38] Their letters reveal, however, that, until the last minute, Smith hesitated whether to go directly to the United States and look there for the types. Robinson, writing from Constantinople on 7 August 1838, did not yet know whether Smith would be sailing with him up the Danube.[39] Fortunately, for Robinson, Smith did accompany him, as it was Smith who wrote to TALVJ in Dresden, asking her to rush to Vienna when Robinson fell dangerously ill.[40] One month later, Robinson, who had recovered, was still in Vienna preparing to leave for Dresden while Smith went on to Leipzig. On 14 November, Robinson was already in Halle.[41] On the way to Europe, Smith used his time to write very long letters concerning issues arising from their journey as well as possibilities for missionary work among the Bedouin tribes in the Sinai and Negev.[42] He continued from Leipzig and in addition to preparing the types and answering Robinson's questions and corresponding with other colleagues, he also found the time for a very long letter starting with: 'I told you in my last, that my next letter might entertain you with a new thing respecting the destruction of Sodom & Gomorrah. So here you have it'.[43]

BIBLICAL RESEARCHES: FROM INCEPTION TO PUBLICATION

The story of Robinson's extended stay in Germany in order to consult with Smith, experts, collections and libraries, while writing his book manuscripts, has been relatively neglected in modern research. One of the primary reasons for this neglect may be that, except for Professors Jay and Hermine Williams, who were greatly interested in Robinson's early years and in his wife's history, no one consulted the vast archival material. Consequently, key documentation was still missing that could supply answers to basic questions relating to the processes of decision-making and book production, including the production of the all-important maps. Carefully deciphering and reading more and more material, we became convinced that the real story of the writing and publishing of the books and maps is actually much more complicated than has hitherto been suggested and accepted in the existing literature to date. In addition to both New England archives, we located illuminating material – mainly correspondence – in various other archives, primarily in Berlin, Gotha and London. This material enabled us to construct an improved narrative of the book production, its background, development and results.

It might be difficult to understand Robinson's obsession with the writing and publication of the book. The description of his intensive work in Berlin and in Halle, for 22 months from November 1838 to 1 September 1840, will not be dealt with here in detail, though it has been only superficially studied and described. The intense correspondence, primarily between Robinson and Smith, sheds much better light on Robinson's work and deserves future study.

The letters do not reveal an early determination to write a book. After reaching Halle in November 1838, Robinson was impressed by the interest demonstrated by his teachers and friends Gesenius, Rödiger and Tholuck, and wrote to Smith that 'they all agree in urging the propriety and importance of handwriting an account of our journey and also the tests of Arabic names [...]'. He decided to take the journals to Berlin in order 'to get the opinion of the geographers there' and accordingly make a final decision.[44]

If Robinson had been eager to publish a book, he very likely discussed it with Smith during the long nights they spent together. Robinson soon wrote that 'in general, our journey & observations seem likely to prove of more value & importance than I at least had anticipated'.[45] He arrived in Berlin on 15 November and wrote, eight days later, that, although most of his time was spent looking for lodgings for his family for the next six months, he had already visited Ritter and Neander, who both offered 'the use of their libraries in the kindest manner'. The dozens of letters that Robinson subsequently mailed to Smith aptly demonstrate Robinson's methodology for studying the material collected, resolving questions and writing a highly accurate text. He began immediately to compare his and Smith's journals, and repeated his positive surprise at their agreement, as 'there is scarcely a difference in the relation of time, except as arising from the difference of our [?]; & no discrepancy as to things seen, etc.'[46]

Lying on Ritter's table, he found Smith's book about his research in Armenia, which Ritter was using for his *Comparative Geography*. Ritter did not waste time and invited Robinson 'to the next meeting of the Geographical Society, which takes place Dec. 8th. If possible, I hope to lay before them a *general* account of our journey, which can then be published perhaps with their sanction.'[47] In June 1839, he was so advanced with his book that he could send the Royal Geographical Society of London (RGS) 'an extract of the journal of travels', published in the *Journal* of 1839.[48]

Working intensively, Robinson continued to collect all possible material, primarily by sending inquiries for additional information to travellers and scholars. He most likely met with Lord Prudhoe in Jerusalem and consulted him on the road from Suez to Wadi Musa in Sinai and received a letter 'with computed distances and the leaving by the Sun'.[49] The reference to Lord Prudhoe, however, is an enigma. Born Algernon Percy, the 4th Duke of Northumberland was a British naval officer and was subsequently, in 1852, appointed First Lord of the Admiralty. In 1816, he had acquired the title Baron Prudhoe, under which he travelled in the East, including Sinai and Egypt, between 1826 and 1829.[50] However, the aforementioned letter is signed 'Prudhoe, Jerusalem, April 15th 1838'; moreover, Prudhoe wrote that he answered an enquiry of 'this morning'! Lord Prudhoe is mentioned in various letters. In 1852, while on his way to his second expedition, Robinson wrote that he could not accept an invitation 'to Syon house from our old acquaintance Lord Prudhoe.'[51] An earlier June 1838 letter states that missionary Whiting 'is 10.000 [Piasters] in debt to Lord Prudhoe's account' and 3.500 to his own.[52] George Backus Whiting and his wife Matilda S. were missionaries for the ABCFM. They arrived in Beirut in May 1830 and were sent to Jerusalem in late 1834 at the end of the rebellion against the Egyptian rulers. They established a mission school but in 1838 'were forced to return to America because of Mrs. Whiting's poor health'. Whiting had been of great help to Robinson and Smith. He is frequently mentioned in the letters in connection with missionary issues, but also with regard to logistical assistance such as hiring horses and supplying money for their expedition. George Whiting returned to Syria in 1840 and died there of cholera eleven years later .[53]

As early as January 1839, Robinson made up his mind about the structure of the planned book. It would have to include important material in four appendices to the main volume: 'Religion Sects, etc., in Palestine', 'Notes on the Arabic Language', 'This Itinerary, etc.', and 'Lists of Arabic Names, etc.'.[54] Soon thereafter, he decided to organise the book according to geographical sections:

> [...]each complete in itself; e.g. from Cairo to Suez, with remarks about Goshen & the Exodus at the close; than from Suez to Sinai, with litr[literary] remarks on the region of Sinai & the convent; thence from Sinai to Akabah; Akabah to Jerusalem, etc. etc. In this way, it will be better to bring in the article which I hope from you on the Religious Sects of Palestine, on a *separate section,* & not as an appendix.[55]

In mid-April, he was able to report, 'I am just now ready to set off from Sinai' and move forward.[56] At the end of 1839, after finishing the long chapter concerning Jerusalem, he wrote, 'I have been able to proceed more rapidly. [...] I am *now* at Jericho; but hope to get back to Jerusalem before the new year. Then there remains [?] the sections to Gaza & Hebron; thence to Wady Mûsa & back, from Hebron to Ramleh & Jerusalem, & there our journey.'[57] On 12 February 1840, he wrote, 'I have really just arrived at Hebron from Gaza'[58] and on 15 April, 'I hope to see the end of my work in June.'[59]

On 15 March 1840, Robinson finished the manuscript. Upon returning to the US in late November, he wrote to Ritter in Berlin complaining about the difficulties he and his family faced in finding a residence and settling down in New York. As for the book, he added that 'the printing of the work is going on in England also', that he was reviewing the sheets in the US and that the American edition 'will be the standard' work. Knowing that Ritter had connections with the King of Prussia, Robinson promised to send them both a copy 'at the very earliest opportunity'.[60]

Only after returning to the US and visiting Boston did Robinson contact his American publisher, Crocker & Brewster. His work was accepted immediately and printing began in New York. The maps were imported from the English edition and met with 'the same degree of unconditional favour'. The American publisher planned 2,500 copies, the London publisher John Murray 1,500 and the Germans only 1,000.[61] *Biblical Researches* was published in English (Boston and London) and in German (Halle), with some differences in the chosen titles,[62] as well as in the exact time of issue.

Both the British and American editions were released in early July 1841, initially without the maps due to some lingering problems. In November, the German edition was completed and Robinson waited expectantly for the scholarly reviews.[63] Indeed, the book was received very favourably. Robinson reported that already by early 1843, 'the London edition (1500) is exhausted & Murray desires to reprint it'. He added that it was still possible to get the American one, which came out later, while the German edition 'was not all printed in November last'.[64] By December of the same year, the stock was entirely depleted and Robinson, upon receiving Ritter's review, complained about the 'neglect & ignorance of such matters which reaches in this country, & especially in this great city of Mammon'.[65]

When preparing to publish the findings from his second expedition, Robinson decided to replace his German editor, the publishing house of the Waisenhaus in Halle (Halle Orphanage), putting his confidence in Ritter's editor, Georg Ernst Reimer from Berlin. Reimer's publishing house, founded by his father Georg Andreas, had extensive and impressive experience in publishing geographical material. Upon visiting the original editor in Halle after returning from the East, Robinson wrote that 'I have seen the *Vorsteher* of the W. H.[Waisen Haus] Buchhandlung, […]; & to my gratification, he makes no claim to the [?]. So that the way is now open for an accord with Mr Reimer.' Robinson left the matter in Ritter's hand, but ensured that 'the Honorar ought this time not to be less than 3 Louis d'ores the *Bogen*; which is the sum fixed by you for the former work'.[66] Reimer agreed to publish the book. As for London, Robinson contracted with Murray to 'publish a new edition of the former, compressed into two vols, […], to which then the new [?] will form a third volume'.[67] The publisher discussed the new edition in letters to Robinson, and added his willingness to prepare the maps for Europe and America, 'if you will only furnish me with the correct drawings'.[68]

The publishing took quite a long time; it seems that the fact that Robinson was not present in Europe was at least partially responsible for the delays. In March 1855, he admitted that 'my work goes on slowly, but thoroughly; & I have much satisfaction and enjoyment in it', adding that 'I am grateful, that my life has been passed to work up my materials; indeed the *Biblical* ground is now pretty much covered; so that should I now be taken away, the work will not be lost to Bibl. Literature.'[69]

In late March 1855, Robinson wrote to Ritter about English proof sheets and translations, which were sent to him 'in order to secure accuracy', also admitting that he was 'continually making additions'.[70] Finally, in July 1856, Murray could write that with the printing finished, he was awaiting the stone plates, which were on their way, probably from the US.[71] The London publisher worried about his timetable versus that of the American publisher and urged Robinson 'that I rely on your not allowing the American publication to take place before the 15th October, at the soonest'. Murray was waiting for the maps, which were, as usual, causing problems and delays, and deciding how many copies to print, 'probably 1000 of vols I & II & 1500 or even 2000 of vol III'.[72] Murray had been a bit pessimistic about the publication date, as by early July he was writing, 'I am glad to hear that the printing of your book on Palestine is at last finished, & that the stoned plates are on their way to me.'[73]

EDWARD ROBINSON AND CARL RITTER

> I cannot let the opportunity pass, […], without giving a renewed expression of grateful & affectionate feeling towards yourself. We often think & speak of you, & see all other daylight the reasons of intercourse with yourself & others, which we enjoyed in Berlin.[74]
>
> The brightest period, connected with my journey to the East, – besides the journey itself, – is the time I spent in Berlin in the preparation of my materials, where I enjoyed such cherished intercourse with yourself & others, […].[75]

Carl Ritter's relationship with Edward Robinson has been described in various studies.[76] They began working together closely in September 1838, when Robinson returned to Germany from his travels in Palestine to begin writing his book. Their relationship was a natural development from previous collaborations and their participation in the *Palästinaforschung*.

Within the 21 volumes of his *Vergleichende Erdkunde,* Ritter dedicated 168 pages to Palestine and Syria without mentioning Sinai even once. Ritter primarily discussed the ancient periods, sacred places and their related traditions, without any treatment of the present and its connection with the past.[77] Subsequently, between 1848 and 1855, he published the four volumes of the *Vergleichende Erdkunde der Sinai-Halbinsel, von Palästina und Syrien,* comprising no fewer than 4,505 pages, out of which 1,141 deal with the Sinai Peninsula.[78] Clearly, no other part of the world merited for Ritter such a detailed study. Scholars maintain that this attention stems from Ritter's strong teleological approach,[79] but, it is argued here, that it would never have reached this intensity but for Robinson's travels, findings and publications.

Robinson, wrote Williams in *Times and Life,* heard Ritter lecturing only coincidently not long before leaving Berlin in 1830:

> I went the other day to hear Ritter the celebrated geographer, he is a very interesting man, more than so, tall and slender with a noble head and forehead, but old-fashioned & plain in his dress & appearance, & reminding much of some vulnerable old New England clergyman. I regret not having found him out sooner, as I should have been much pleased to have made his personal acquaintance. He is also a pious man, tho' he is much devoted to his scientific pursuits, that some of them complain here, that he is not open & pious enough.[80]

Robinson most likely attended lectures given by others in geography and in what both fathers of modern geography coined 'Comparative Geography'. One of the lecturers was Alexander von Humboldt, who had begun teaching in the university, which at the time was named for King Frederick William III, in 1827.

In his book, Williams touches upon Robinson's cartographic work and Ritter's role in that work only briefly. At the end of the discussion concerning the *Biblical Researches* Williams wrote:

> Perhaps, Robinson's most important accomplishment, however, was the drawing of new maps of the whole region. For this purpose he hired a young German map maker, Heinrich Kiepert, [...]. It was an amazing and fruitful response for his teacher's [Moses Stuart, H.G.] cry for 'Maps, Maps, Maps'. Through his efforts the maps of ancient Israel were thoroughly revised and improved; modern cartography of the Holy Land begun.

> During this period he consulted not only his old teacher, Gesenius, but geographers like Carl Ritter and Alexander von Humboldt and philologists like Roediger. He also read before the Geographical Society of Berlin [...][81]

Williams added that, during the 1850s, particularly given the death of Gesenius in 1842, Ritter became 'Robinson's major correspondent in the continent'. Robinson, who had all the understanding for the latter's 'great pressure of labour which has been upon you for the past year. A volume of 1100 pages a year, on even every two years, is no trifle, especially when all your other [...] are taken into the account', continued leaning on him as his main collaborator in Germany.[82]

That Robinson would dedicate the German edition, 'corrected if not wholly translated by Therese', to Carl Ritter[83] over any of his other teachers in Germany is not obvious. Robinson dedicated the American edition to Moses Stuart, undoubtedly his most influential mentor, responsible for his strong, scientific scriptural approach and for his studies in Germany. Clearly, Ritter's influence was equally significant to Robinson's academic work. It is then not surprising that contemporary German scholars such as Gotthilf Heinrich von Schubert saw in this classic work 'a property of our homeland and our mother-language'.[84] But of all the reviews written by German scholars, it seems that the reviewer in John Murray's London *Quarterly Review* should be cited:

> We opened this work with a feeling of weary despondency at the prospect of three more volumes of Travels in Palestine; we closed them with respect and gratitude to the author, not unmingled with a little blameless national jealousy. We are not altogether pleased that for the best and most copious work on the geography and antiquities of the Holy Land, though written in English, we should be indebted to an American divine.[85]

There is little doubt among scholars that Robinson adopted many of Ritter's ideas. They maintain that Robinson's highest ambition was the explanation and illumination of the Holy Bible;[86] his goal was to investigate the relatively well-known historical country of the Holy Scriptures, the Holy Land, between the Mediterranean and the Jordan, from Dan to Beer-Sheva.[87] Robinson accepted Ritter's teleological as well as deterministic perceptions. They both argued for a deterministic approach to the study of the country, one that requires knowledge of the mutual relations between its physical characteristics and historical development. Accordingly, Robinson paid special attention in his books to the morphological outlines and planned (but did not complete) a two-volume study summarising the physical geography, topographical history and geographical history of the 'Holy Land' and its neighbouring regions.[88]

Ritter and Robinson saw the uniqueness of the country in the historical periods, in its position in the Jewish and, especially, the Christian religions. The richness of evidence in the field that could verify religious and other historical sources led them to concentrate almost exclusively on the past. When they noted aspects of the present country, it was only for the purpose of explaining the past. This approach, further developed by leading Scriptural Geography scholars such as the Scottish theologian George Adam Smith, has been discussed in numerous publications.[89]

Ritter's influence was not completely one-sided. On the contrary, Robinson, with his unique philological approach, field research in the Holy Land and publications, made a strong impression on Ritter's thought, ideology and works. One can conclude, therefore, that the fact that Ritter's intensive study of the Holy Land started only in the early 1840s is not mere coincidence. It probably has to do, more than anything else, with Robinson's work and his close relationship to Ritter. To determine the depth and the nature of this influence, further study is required.

To say that Ritter used much of Robinson's material in his four volumes of the *Comparative Geography of Palestine* is a complete understatement. Volume II includes a long and detailed 'Review of the Authorities on the Geography of Palestine'. Ritter affirmed in the early 1840s that Berlin's Geographical Society under his leadership was the first to acknowledge the great importance of Robinson's work and to publish it.[90] When mentioning Robinson's book in his *Vergleichende Erdkunde*, he frequently used terms such as 'admirable' and 'classic'. It is essential to cite here long sections of Ritter's assessment of Robinson's and, we should not forget, also Smith's work, as translated into English by Ritter's Scottish student William Leonard Gage:

> The union of that very close observation of the topographical features of the country which characterizes the work of Burckhardt, with many preparatory studies, particularly with a thorough familiarity with the Bible, and with philological and historical criticism, and the thorough acquaintance with the colloquial language of the country enjoyed by Mr Smith [...], make this work, prepared as it was after the severest toil, a classic in its own field, a production which has already set the geography of the Holy Land on a more fixed basis than it had ever had before, and which will ensure its continued advance. No previous work had collected a greater store of new and important discoveries of a historico-critical character, says the competent judge, J. Olshausen;[91] and the admirable principles of investigation which are unfolded in Robinson's work, will serve as a beacon for all future explorers, *who shall endeavor to read the word of God by the light reflected from the scenes amid which it was recorded* [my emphasis].[92]

Ritter continued with what seems to be almost exaggerated praise in response, perhaps, to critical scholars:

> The work has marked an epoch in biblical geography. The universally recognized merits of its author – who has been, as becomes a true scholar, not grudging in his commendation of worthy predecessors in the same field, but who has had at times, in his eager search

> after truth, to be the open foe of convent legends, the light tales of tradition, and the gross historical errors which lay in his way – have not prevented his being attacked by all kinds of adversaries, some of them men of superficial attainments, some of them men actuated by base motives or by passionate animosity.[93]

Robinson did not set out to publish a book with no errors, Ritter continued, and hence has not been engaged in defending his opinions. Rather, he perceived his work as the basis for more advanced and accurate studies:

> The task of his life, first to last, lay before him rather than behind him; and the German editor of his later researches (Rödiger) very justly says that Robinson's greatest merit lies in his kindling into life that great interest in the topography of the Bible scenes, which has prompted a very high class of minds to explore the region with exhaustive skill [...][94]

Emil Rödiger, German orientalist from Halle, was one of three German colleagues in addition to Ritter with whom Robinson consulted while writing his books. Rödiger also worked closely with Smith in preparing the Arabic types, exchanging frequent letters with both. 'The orientalists here, Gesenius, Roediger, & Tholuck, received me with open arms,' wrote Robinson to Smith.[95] Rödiger connected Robinson with the Halle Waisenhaus publishing house, where he published his German edition.[96] With Gesenius, who established Hebrew lexicography in Germany, Robinson developed a very close relationship.[97] The leading theologian August Tholuck, who from 1826 lectured in Halle on the Old and New Testaments, influenced Robinson greatly through his teaching and preaching.[98] All three figures are mentioned frequently in the letters and conducted an intensive letter exchange with Robinson.

Ritter wrote a long and detailed description of Robinson and Smith's working methodology, the ideology and principles behind it:

> The two fundamental principles which Robinson and Smith have laid down for their guidance in determining the historical value of the traditions of Palestine, were these, that different weight is to be attached – (1) to the later traditions which have arisen since Constantine's time, and which, springing from the changed ecclesiastical condition of the land, have been largely diffused by those who were not the primeval inhabitants of the country, but resident aliens, so to speak; and (2) to the primitive and indigenous traditions, rooted deeply in the Semitic character, living in the mouths of the common people, and perpetuating themselves in the local names of places, since the Arabic now spoken is so akin in its general features to the Hebrew which it has supplanted, that it changes but slightly the old words, and leaves the roots visible; [...]. These indigenous words were never regarded as important by the Byzantine ecclesiastical

> authorities; nor were they observed by the earlier travellers, who surrendered themselves unreservedly to the guidance of monks, and contentedly received whatever they told them. But the more ancient tradition both Robinson and Smith found never to deceive them [...][99]

Ritter had been quite quick to show his appreciation of Robinson and Smith's work and did his best to help them prepare and publish the book. In early February 1839, Robinson wrote that he 'continues his kindness & attention; & has recently made me a present of a fine copy of the great French maps of Palestine, taken by the great map-maker Grimm' (see Fig. 1.2).[100] In another, very long letter in German, dated March 1850, which serves as a summary of their scientific relationship and collaboration, Ritter wrote that Robinson can now see in the second half of his *Palästina* a testament to the rich treasures he found in the *Biblical Researches*.[101] In expectation of the third volume, Ritter wrote again praising Robinson's contributions:

> for which I thank you [...] Without them it would have been impossible for me to write my works and to reach such a scientific standard. [...] Without your guidance, it would have been totally impossible to start on this way, [...] each page gives evidence for it.[102]

Both Robinson's and Ritter's books were considered by contemporary scholars as a 'must' for whosoever plans to study the country. 'Too many books are, when travelling anywhere, but especially in Palestine, a burthen', wrote the Dutch cartographer Charles William Meredith van de Velde, but Ritter's *Erdkunde* and Robinson's *Biblical Researches* 'I thought, however, indispensable'.[103]

Robinson and Ritter remained close until the latter's death in 1859, only four years before Robinson himself died. They maintained a relatively intensive correspondence and sent each other new publications punctiliously, of course, primarily their own. They also shared a group of friends and colleagues, including John Washington in London and the cartographer Heinrich Mahlmann in Berlin.[104] In this way, they kept abreast of all new developments in the field to which they were dedicated, the study of the Holy Land. In July 1842, Robinson wrote that he had heard from Smith about two new developments: Robe's map of the sources of the Jordan and two printed reports by ABCFM missionary Samuel Wolcott. Naturally, Smith met both Robe and Wolcott in Lebanon.[105] Major Frederick Holt Robe's sketch was immediately forwarded to Mahlmann for correction of the existing maps and from him to Ritter, who published it that very same year in the *Monatsberichte über die Verhandlungen der Gesellschaft für Erdkunde zu Berlin* (*MVGEB*), the periodical of Berlin's Geographical Society. Robinson was forced to wait about a year to publish the map in the first volume of the *Bibliotheca Sacra*, of which he was the editor and primary contributor.[106] Known later for his hymn writing, Wolcott, a graduate of Yale and Andover, arrived in Beirut in 1840 (in 1839, according to Robinson) and served in Syria for two years. Robinson described him as 'an active and intelligent observer of men and things', and included his communications in the very first volume of *Bibliotheca Sacra*. There he published Wolcott's letter to Smith, dated 'Sebbeh, Masada, March 14th, 1842'.[107] His two papers concerning the topography of Jerusalem and 'excursion to Sabbeh on the W. coast of the Dead Sea, which is now fully identified with Massada', were published in 1846 in the *Bibliotheca Sacra and Theological Review*, the periodical replacing Robinson's *Bibliotheca Sacra*, which appeared only once, in 1843.[108] Another prominent example of keeping abreast of new developments can be seen in Robinson's detailed report of his discussion with scholars in England, most notably with George Williams, who 'set on [...] a crusade in behalf of tradition & the Holy Sepulchre'.[109] Robinson had also to deal with the fact that Williams was joined by Ernst Gustav Schultz, who from 1842 served as the first Prussian consul to Jerusalem and 'whose authority will naturally be such greater than that of Mr Williams'.[110]

Given their relationship and scientific collaboration, it is no surprise that Ritter was the first scholar who Robinson consulted when he prepared for his second expedition. Robinson sent a detailed letter containing his timetable from Stralsund (in Pomerania) on 8 July 1851, directly after arriving in Germany.[111] The two met in Berlin and then Robinson continued on his travels and sent another letter to Ritter:

> My main object in now writing is to ask, (which I forgot to do in Berlin,) that you will be so good as to note down all points of inquiry which may arise, & which may desire [?] to need enhanced [?] examination on the spot; – so that I may have the benefit on my return in Jan[uar]y [1852, planning to Palestine]. I have already seen such points; but many more will doubtless arise to you.[112]

'Besides you will have the advantage in Berlin of talking over the ground with Professor Carl Ritter and next to yourself he is best acquainted with the subject,' wrote Washington to Robinson in 1852, upon hearing that Robinson was embarking on a second expedition. This sentence summarises the contemporary perception of the Scriptural Geographers and their contribution to the study of the Holy Land.[113] The question remains as to whether Ritter ever visited the Holy Land. According to his own testimony in a letter to Robinson, Ritter's advanced age of 72 years prevented him from joining Robinson on his second visit.[114]

CARTOGRAPHY: HEINRICH BERGHAUS

Scholars agree that Ritter played a major role in the development of the cartography of Palestine. His methodological approach included an extensive use of maps in his publications.[115] Hence, some of the leading cartographers in Berlin, who were drawing his maps regularly, began working as well on the two-dimensional description of the Holy

Land. Ritter introduced other scholars of Palestine to 'his' cartographers, which resulted in some of the most advanced maps of the region, the country and Jerusalem drawn by professionals such as Julius Ludwig Grimm, who was born in 1806 and died suddenly in 1834, Heinrich Berghaus and Heinrich Kiepert. Ritter gave the cartographers access to all the data he accumulated.[116] He also worked closely with them, as demonstrated by his correspondence with Berghaus.[117] In his 1835 *Memoir to the Map of Syria* (see Plate 1), Berghaus described the works of over thirty travellers and explorers provided by Ritter. Berghaus also began working with Edward Robinson, but being too busy, at least according to accepted sources, he was replaced with Kiepert, who then launched a career of nearly forty years of cartographical work on Syria and Palestine.[118]

Robinson's most important accomplishment, according to Jay G. Williams, was the drawing of new maps of the entire region. For this purpose, he hired Kiepert, who detailed, in Appendix B of Volume III of the *Biblical Researches*, the various sources used and how Robinson engaged himself in the minute details of their work together. The accomplishment was an amazing and fruitful response to his teacher Moses Stuart's preaching for 'Maps, Maps, Maps'.[119] Hence, it may well be that Robinson took this task upon himself even before arriving in Germany. What is certain, however, is that through his efforts the maps of ancient Israel were thoroughly revised and improved: modern cartography of the Holy Land had begun.

Cartographers were clearly attracted to the work of Robinson and Smith and preferred their data to that of others. Heinrich Berghaus, who was one of the first to acknowledge the exceptional value of their research, wrote to John Washington in April 1839:

> After having studied many documents serving geographical, especially cartographical purposes I convinced myself that among all those who ever since Niebuhr have been travelling over the eastern countries it is our compatriot Burckhardt who deserves the highest praise on grounds of his meticulous attention to even the most unobtrusive phenomena and his precision in trigonometrical and time measuring when determining distances. Even if I expressed myself in different words, I gave public and unreserved utterance to this conviction in my memoirs on my maps of Asia.
>
> However, at present I see myself under the necessity to considerably modify this opinion after having perused the diaries of Messrs Robinson and Smith, written during their travels. [...] The observations made by the aforementioned travellers being so vast and comprehensive, their description of the shape and relief of the country so exact and beyond any doubt that they will enable the geographer to design a special map which will leave little to be desired.[120]

This quotation is only slightly different from what Berghaus wrote to Robinson only a week after receiving from the latter 'my itinerary & topographical notes as far as Jerusalem', which he wanted to examine for a map he was making. The fact that both Berghaus and

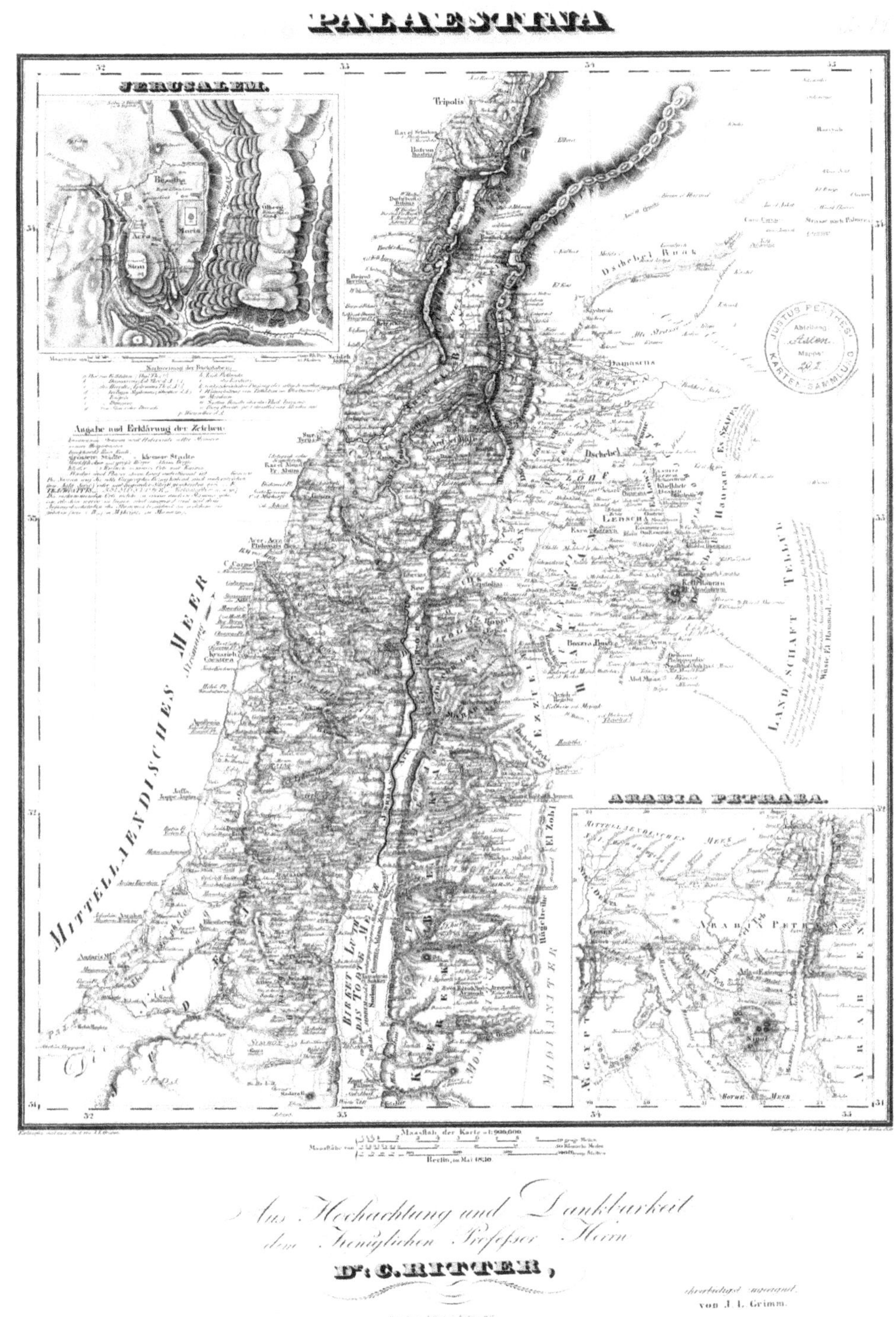

Figure 1.2 Julius Ludwig Grimm, *Palaestina* (Palestine) (Berlin, 1830)
Source: FBG SPK 547-112620698

Ritter praised the Swiss explorer Johann Ludwig Burckhardt[121] as presenting even better than this 'example of an explorer', is the best proof of the esteem with which they held Robinson.

Berghaus's enthusiastic commendation impelled Robinson to go forward with publishing a book and maps. Robinson wrote to Smith:

> Such a testimony, wholly disinterested, & from the person perhaps best qualified of all men living to form a correct judgment on the subject was to me as gratifying as it was unexpected; & I send it to you, that you too may see & feel, that our labours are not likely to prove in vain, nor without acknowledgement.[122]

Heinrich Berghaus was active in many geographical and cartographical enterprises in the nineteenth century. He was co-founder of the Geographical Society in Berlin (1828), founder (1839) and manager of the *Geographische Kunstschule* in Potsdam and a close friend and house cartographer of Alexander von Humboldt and Carl Ritter.[123] This widespread and intensive cartographic activity included his epoch-making *Physical Atlas* published in Gotha since 1838 and in an English edition since 1843,[124] and cartography of the Near East. The latter was of great importance for the development of the region's professional cartography, primarily with regard to field observation and standard of compilation. Berghaus made use of all available material, maps, measurements and written and oral descriptions in order to compile a map that met the developing criteria of modern cartography. This is apparent from his highly instructive *Memoir to the Map of Syria* (1835), where he notes that he needed the reports of eight travellers in order to establish the distance between Jaffa and Jerusalem. Carl Ritter announced that 'his [Berghaus's] work opened a new era in the cartography of Eretz Israel and Syria',[125] and van de Velde praised his map, stating that 'it will at all times remain a monument of indefatigable zeal and extraordinary discernment'.[126]

In actuality, Berghaus was the primary compiler of only this one pioneering map of the Holy Land. The map, containing two parts with different scales (1:450,000 for most of the country, 1:800,000 for Arabia Petraea), was published in 1835 as part of Berghaus's *Atlas von Asien*. Berghaus dedicated it to two leading Palestine explorers, Colonel Pierre Jacotin, who at some stage commanded the French military cartographers during Napoleon's invasion of Egypt and Palestine and in 1810 published the atlas under his name, and to the 'explorer in disguise' Johann Ludwig Burckhardt.[127]

In his letter to Smith dated 22 January 1839, Robinson summarised his consultations and arrangements with Berghaus concerning the maps. Robinson sent examples to Berghaus, who reacted enthusiastically, and they met again. Berghaus declared that 'his map of Palestine & all others are now without value', and offered 'to construct two large maps on the same scale as the former, to include the whole territory between Beirût & Sinai, constructed after our notes'. Robinson, who had already consulted with Ritter on the issue, told Berghaus that he is 'ready to give him all the maps & plans for the Book.' Berghaus agreed and, optimistic, Robinson added that 'something will come of it.'[128]

They did, however, expect difficulties with their plan. The main one was financial, given the cost of engraving such large maps and the comparatively narrow distribution and sale anticipated for them in Germany. Though German, Berghaus preferred the 'patronage & aid of the Geog. Societ. of London.'[129] He proposed applying for the Society's patronage in exchange for an agreed number of copies. As an inducement, he suggested preparing the maps in English. Ritter was favourable and Robinson was quite sure that both Ritter and Humboldt would help. Berghaus proposed introducing Robinson and Smith to the Geographical Society and its secretary and journal editor, British military officer and geographer John Washington, by sending 'a copy of his map of our route across the desert etc., for their journal,' together with a short memoir.[130]

Berghaus's goal in this small map was to illustrate the pioneering route of Robinson and Smith through the little known western Negev, the southern part of today's Israel.[131] The map was intended to illustrate Robinson's report, one of the first published fruits of the expedition. He compiled the map and submitted it to Washington for publication in the *Journal of the Royal Geographical Society of London* (*JRGS*). Its publication in the 1839 volume had unforeseen and negative consequences.[132] The small and relatively unknown map, titled *Part of Arabia Petraea and Palestine*, turned out to be of exceptional importance for the scientific study of the region, being the first to provide a significant description of the area. Surprisingly, the published map also included the addition of routes of another explorer, the French Count Jules de Bertou. These were of the eastern Negev and followed de Bertou's text published in the same *JRGS* volume. Also included was a small inset of the Lebanese city of Tyre, preceded by another publication of de Bertou.[133]

Berghaus had not agreed to this British addition to his map. He did not even know about it until Robinson received copies from the RGS after publication. Robinson complained bitterly that 'they have taken the liberty to add Bertou's routes from Hebron & back, as if put on by Berghaus, they are implying to the journal readers that there is some connexion between us. This I regret exceedingly; for his specifications are false in several respects.'[134] The entire incident sheds an interesting light on scientific connections and the ethics of international knowledge transfer during the first half of the nineteenth century.[135]

The production of this pioneering map was part of a drama that led Berghaus to stop cooperating with the RGS and with Edward Robinson. John Washington was certainly one of the central personalities in the world of geographical research and exploration during the mid-nineteenth century. After a naval career of nearly twenty-five years, in 1836 he was appointed secretary of the RGS, of which he was also a co-founder. Then, as he wrote to Robinson in 1841, he returned to the navy 'to resume the more active duties of my profession'. He went back to sea executing coastal surveys and primarily intelligence missions until 1855, when he replaced the legendary Francis Beaufort as Chief Hydrographer of the Navy.[136] Washington maintained strong connections with leading geographers and cartographers of the time, exchanged materials with many of them and during mutual visits met both 'founders of modern geography', Humboldt and 'our excellent friend Professor Ritter'.[137] He even promised to tighten the connections between them in a letter to Berghaus in September 1837.[138] The British and Prussian geographic and cartographic worlds of that period were strongly connected, engaging in intensive discourse followed by exchange of material, data and technology. Ritter and the German cartographers regularly exchanged materials with the Hydrographic Office of the Navy under Beaufort.[139]

Naturally, Berghaus was the first cartographer with whom Robinson consulted soon after settling in Berlin to write his book and it seems that both were expecting a long-term working relationship. Robinson elaborated on his expectation in an early letter to Smith:

> On Monday last I went again in Potsdam to see Berghaus. I had sent him some of our bearings, on which he had constructed triangles including the Mount of Olives, Mar Elias, Bethlehem & the Frank Mountain. His remarks were, 'that the results showed a wonderful degree of accuracy considering the [?]'. He gave me some important suggestions as to the arrangement of the bearings & of the information necessary for the construction of a new map, which he seems quite disposed to undertake of his own accord. [...] Ritter says no person can do it so able as Berghaus, under my supervision.[140]

Robinson regularly mentioned Berghaus in his letters to Smith, referring to him as the best cartographic authority. Kiepert, however, is mentioned for the first time only in a letter dated 24 December 1839, more than a year after Robinson began to work on his book.[141] An early and important suggestion of Berghaus, which Robinson immediately adopted and executed, was to construct a detailed time-distance table of the expedition. This turned out to be crucial for the compilation of the maps. Not long after beginning his work with Robinson, Berghaus very quickly understood, and admitted to Washington, that, although only four years had passed since the publication of his large map of Palestine, a new map was needed that would contain all the corrections and new facts gleaned from the American's new itinerary. In the same letter to Washington he added:

> I am of opinion, that it would be *of a great loss for geography*, were the materials collected by Messrs. Robinson and Smith not to be used for the construction of a map on a large scale. Their Journey, [...] would be deprived of its finest fruits, [...] should the results of their measurements and observations be published only in the form of a journal.[142]

Robinson compiled the time-distance table in about two weeks. In his very next letter to Smith he wrote that, 'for the last fortnight I have been busily occupied in drawing out from our Journals, a full itinerary, of stations, courses, distances, bearings, with topographical notices, all in short that can contribute to the construction of a new map'. Completing it as far as Jerusalem, he took it to Berghaus, 'who seems anxious to make the construction' as 'he seems to prize it very highly'. Ritter suggested printing the table as an appendix to the planned book.[143]

In March 1839, Robinson was able to report to Smith that the big map was in progress:

> in constructing the net of triangles, Prof. Berghaus starts from the base I measured at Jerusalem & [on] the Mount of Olives; & does not take any note of our routes & courses; partly because of the uncertainty of the *hour* with horse & mules both in itself, & partly on account of the ground travelled over.[144]

As mentioned above, claiming that such works are not encouraged in Germany, Berghaus asked the RGS directly 'for the construction and publication of a *Special Map of Palestine and Arabia Petræa in the English Language*.'[145] Robinson joined him enthusiastically and the project developed in an intensive letter-exchange between Berlin/Halle and London. In a letter from Robinson dated 4 June 1839, he specified the details of the planned map including its borders and noted that it had to be on a scale of 1:250,000 and engraved on two sheets. 'I hope,' he added, that 'Prof. Berghaus will be able to make out a far better map than has yet appeared.'[146] Only one day later Berghaus sent another letter to Washington, this time in French, a language he probably controlled better than English. He mentioned numerous studies that he would like to have in order to complete his compilation. Included were studies of Lord Lindsay, George Henry Moore and William G. Beek, who were the first to publish the fact that the Dead Sea is lower than the Mediterranean, Beaufort's collected maps in the 'Hydrographic Office of the Navy', Francis Rawdon Chesney and his 'Euphrates Expedition' and the travel-narratives of the British officers Charles Leonard Irby and James Mangles.[147]

However, Smith, who stopped in London to meet with Washington on his way from Leipzig to the United States, alerted Robinson to developing problems: 'Cap Washington manifests much interests our investigation, but the project, as Berghaus suggested, is much too big and expensive and will not find enough subscribers.' Smith was also still very angry about the addition of Bertou's route to the Negev map and indicated his desire to meet again with Washington to clarify the story.[148]

Three months later, Robinson also expressed his disappointment. On 18 September 1839, he wrote that 'my maps, I regret to say, do not make good prospect', assuming that Berghaus was simply too busy.[149] Gerhard Engelmann maintains that Berghaus had been busy building his *Geographische Kunstschule*, as he received permission for the school on 14 July 1838 and its inauguration took place on 1 April 1839.[150]

This may not be true, or may be only a partial reason for Berghaus's retreat from working with Robinson. In his letter dated 5 July 1839, RGS Secretary Washington politely rejected Berghaus's request for help in financing this new map of Syria:

> Both your letter & that of Dr. Robertson [sic; he means Robinson] were duly laid before our council, and I am desired to express to you their best thanks for the details respecting your proposal of Syria, and to say that the numerous calls on the funds of the Society, having no less than three foreign exploring expeditions on hand (Kurdistan, Guyana, and White Nile) [...], prevent them from acceding to your request any further than by subscribing for copies of the intended map, on the understanding that it will not exceed in price £1 sterling for a copy, [...][151]

Berghaus was not expecting a rejection from the RGS, a society that he praised and adored. This answer was part of a longer letter responding to Berghaus's letter dated 30 April 1839, in which, in addition to the financing request, he raised twelve questions that he thought might help him with the planned map.[152] Washington did answer them in detail while pointing out that RGS did not possess any documents concerning the region; his information came from his habit of interviewing travellers who visited the society.[153] It seems, however, that Washington's answers were no longer of use to Berghaus. Consequently, we find that Robinson wrote to Smith in late 1839 expressing his view that Berghaus had stopped cooperating due to this negative response from Washington for a London Society grant. Robinson had to start looking for another cartographer:

> My maps alas! Get on less well. You are aware [that] Berghaus undertook them, & was to have them done by July 1st. [It] went well enough till he was disappointed in aid for his grant from the London Society; & after that he D't within. I had been warned of his character, & that there might be difficulty [?] my materials; [...]. On enquiring on him in September what I might expect the maps, I could at first get no reply; & at last another indefinite reply, that he could do nothing before November.[154]

CARTOGRAPHY: HEINRICH KIEPERT

Born in Berlin, Heinrich Kiepert (see Fig. 1.3) studied history, archaeology and antiquity studies between 1836 and 1840 at its young university and eventually found his way to Carl Ritter's lectures, which had the strongest influence on him.[155] Kiepert did not join Berghaus's *Geographische Kunstschule*, founded in Potsdam in 1839; he developed all his cartographical skills at the university. His collaboration with his mentor lasted until Ritter's death in 1859. Kiepert published his first maps, dedicated to the classical world, from 1835 and his first Palestine maps, the fruit of his work with Robinson, in 1840 (see Plates 2, 3 and 4).[156]

It was Ritter, who introduced Robinson to Kiepert. Upon realising that Berghaus would probably never produce the maps, or at least not on time, Robinson turned again to Ritter, who advised him to approach the young cartographer. Kiepert met with Robinson, looked at the materials and, although very young and inexperienced, declared he promised Robinson

Figure 1.3 Heinrich Kiepert, photography, 1891
Source: Leibniz Institute for Regional Geography, Archive for Geography, Por-Kiepert01

to meet his timetable.[157] By February 1840 Kiepert could do the work, demanding an advance on his payment. Robinson had finished the manuscript and the book was being prepared for publication, but he could only write that the maps 'are not yet ready, but are a work in progress'. With Kiepert in Berlin, Robinson had the additional advantage of being able to supervise him directly.[158] In mid-April, Robinson wrote that 'the maps are coming along well. Kiepert, though young, is far more thorough & conscientious than Berghaus; & in this way has made order out of our material where B. found only confusion' Robinson also detailed the maps planned:[159]

> Sinai & Southern Palestine, as far as Hebron.
>
> Palestine from [Hebron?] to Tyre on a scale larger than Berghaus' large map, with a side map extending to Beirut & Damascus.
>
> Plan of Jerusalem, corrected from Catherwood's & extending N. to the tombs of the Judges, & south to the lower bend of the Kidron.[160]
>
> The environs of Jerusalem on a scale 4 times greater than Berghaus' large map.

In October 1840, while in London on his way home, Robinson could at last write to Smith that 'the maps [...] are at length completed. [...] They have also been engraved in Berlin on stone, by a very faithful & intelligent man, Mr. M[ahlmann].' Robinson brought three of the finished maps ('The plan of Sinai, of Jerusalem & environs, & the southern section of Palestine', see Fig. 1.4) with him to London, presenting them before the British artillery officer William Martin Leake, Washington and others. The remaining maps were sent after him.[161]

The first large map, drawn on a scale of 1:400,000, had been edited by Ritter and was based on the maps of the Holy Land drawn by Grimm and Berghaus. The latter added a memoir, highly important for Kiepert as well as for present day scholars of the history of Palestine's cartography. In his own memoir, Kiepert confined himself to narrower issues such as astronomically fixed points, as was his custom 'after the example of similar memoirs' as most of the material was included in the 'very valuable' memoir of Berghaus. Kiepert added the data meticulously collected by Robinson and Smith. He considered the routes of Robinson and Smith the highest authority, 'which in [their] minute specification leave far behind them the reports of all other oriental travellers'.[162] For southern Lebanon, an area that Robinson and Smith did not explore fully until their second visit in 1852, Kiepert used a sketch by Isaac Bird. Bird, like Smith, was an ABCFM missionary, a Yale graduate who had settled in Beirut in 1823. Unfortunately, his sketch, as well as Ehrenberg's,[163] is under 'war loss' (*Kriegsverlust*) in the map department of the Berlin State Library.[164] According to Ritter, Kiepert's turned out to be the best and most detailed map to date.[165]

Neither Robinson nor Kiepert had any problem locating sources they needed between Ritter's library and his geographical society. In 1978, the *Geographische Zeitschrift* published a special issue of more than 450 pages with the bibliography of Ritter's library and map collection auctioned in 1861 by Theodor Oswald Weigel in Leipzig:[166]

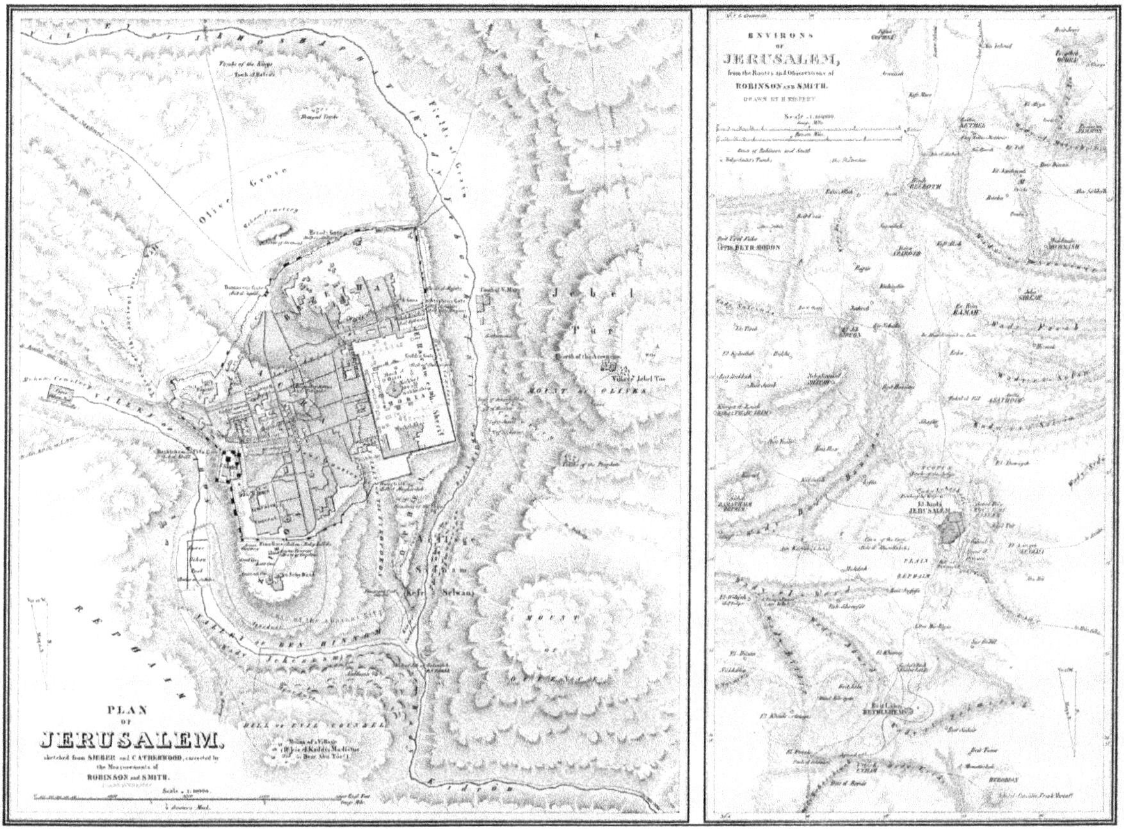

Figure 1.4 Heinrich Kiepert, *Plan of Jerusalem sketched from Sieber and Catherwood, corrected by the Measurements of Robinson and Smith* (Berlin, 1841)
Source: SBB-PK Kart. X 1401

> Originally Carl Ritter's legacy included 12,000 maps, 18,000 books, 484 notebooks with collections, more than 1,000 letters as well as travellers' reports collected by Ritter [...]. Then the *Königliche Bibliothek* bought 2,554 map-leafs [...]. In 1867 the library bought from the firm Weigel 17 volumes (or capsules) with letters to Carl Ritter, collections and the manuscript of the '*Erdkunde*' and, in addition, 475 books.[167]

It is impossible to describe the collection, as mere numbers are inadequate for creating the right impression. Under 'Syria' are 'only' 37 publications, but these are followed by 'History of the Jews, Palestine, Phoenicians, Philistines' with 120 titles.[168] The collection also includes more than 150 maps of Asia.[169]

The principal strengths of Kiepert, Ritter's student and successor, were described accurately by his contemporary, Carl Sandreczki, a Bavarian-born missionary and physician who was also involved in the scientific study of Palestine. Sandreczki wrote that it was very unfortunate that Kiepert was only able to stay in Palestine for such a short time (on his

1870 visit), because his knowledge of the research questions and ability to evaluate prior findings and trace what had not yet been discovered was unmatched.[170]

Collaborating with Robinson marked the scientific development in Kiepert's career as a young and promising cartographer. While compiling the early maps, Kiepert encountered the primary difficulty that accompanied all his work on the Holy Land, namely the need to compile maps based on sporadic descriptions including historical sources, traveller and pilgrim itineraries and random measurements, in addition to the relatively accurate geographical data attained by professionals. He adopted a highly critical approach that involved using the most up-to-date scientific instruments together with locating and the critical study of as many sources as possible, which he compared meticulously to Robinson's results. In addition, the various sources contained inaccurate transcriptions and pronunciations of Arabic. For this discipline, the skills of Eli Smith were critical. These challenges prompted Kiepert to continually develop his interests and expertise. He acquired a deep and thorough knowledge of historical as well as contemporary sources, was quite pedantic in staying current with geographical research and, over time, achieved a good knowledge of many European and Oriental languages, including Hebrew, Arabic, Armenian and Turkish.[171]

Kiepert made the most of Ritter's Berlin, which had developed into a centre for Palestine studies, probably the largest and best informed in Europe. Naturally, this scholarship and plethora of sources were available to him.[172] Kiepert's detailed memoir testifies to the difficulties he encountered in his cartographical work. In his memoir he claims that he could not rely on contemporary calculations and triangulations, as they turned out to be insufficient. Even some of Robinson's calculations and triangulations, which were considered better than Burckhardt's, particularly when used in conjunction with Jacotin's and Moresby's measurements, required extensive research in order to calculate the routes. Moreover, it is fascinating to note that Kiepert did not hesitate to criticise aspects of Berghaus's work.[173]

In spite of Robinson and Smith's remarkable fieldwork and endless efforts to produce the most accurate data, Kiepert knew that, in order to compile and draw the best scientific map, he needed additional sources. This is natural and understandable; what is surprising is the speed with which he developed his expertise and rank as a leading Holy Land cartographer, independent of Berghaus.

The results of Kiepert's collaboration with Robinson are very impressive. In his bibliography of 'Palestine literature' encompassing publications up to 1878, Reinhold Röhricht listed eighteen Kiepert maps of Palestine and sections of the country beginning in 1840, some published in several editions.[174] In his 1999 bibliography of Kiepert's cartographic publications, Dirk Hänsgen enumerated more than forty items dedicated to the Holy Land and its neighbouring countries, published until Kiepert's death in 1899.[175] 'Next to Berghaus and van de Velde, he easily takes first place among the Palestine cartographers,' wrote Titus Tobler, the highly critical scholar titled 'father of the German study of Palestine', in his pioneering bibliography of Palestine literature published in 1869.[176] It is only natural that Ritter praised Kiepert's work, but it nevertheless seems important to cite his comments:

> The maps, which were constructed with the rare skill of Dr Kiepert from the voluminous data furnished by Robinson, the result of his innumerable measurements, and which were lithographed in the most faithful and beautiful manner by H. Mahlmann, raised the cartography of Palestine one level higher than even Berghaus reached; and they remain perhaps the very finest efforts of skill which have appeared either in or out of Germany [...].[177]
>
> [...] that it was necessary to construct a new and independent map of Palestine, which should, so far as the eastern shore of the Jordan is concerned, have done little more than repeat what Berghaus had already compiled, but which in all that makes up Palestine proper, should be an original work. This task, which was to illustrate Robinson's 'Biblical Researches', was accomplished by H. Kiepert in so masterly a manner, and in every respect with so thoroughly scientific a spirit, as to win the applause of all scientific judges, and to be the model for all subsequent works of its kind.[178]

Scholars and travellers of the Holy Land made immediate use of Kiepert's maps. Samuel Wolcott used the map for his 1843 expedition. While inspecting the vast area of the Dead Sea from the cliffs of Masada, he was deeply surprised and impressed: 'it is certainly surprising that it should have been left for American research, at so late a period, to define with any degree of correctness the shape of this singular sheet of water; as it is gratifying that it has finally been done so accurately.'[179] Scholars who planned research expeditions in Asia Minor and Palestine went to Berlin to consult with Kiepert and request cartographic instructions. In this respect, too, Kiepert followed Ritter, becoming a source of data, guidance and publication assistance for Holy Land explorers. Kiepert's instructions usually included maps and sketches depicting the regions that he thought needed further exploration, as well as detailed instructions for field study.[180]

Both the biblical Holy Land and contemporary Palestine were central to Kiepert's scientific and cartographic activities. Indeed, as noted above, Röhricht cited eighteen maps depicting different parts of the Holy Land that were produced over five decades of extensive work. The maps can be sorted into two major categories: contemporary Palestine, compiled from travel diaries and primary and secondary literature, and biblical Palestine. One can divide these further into subcategories sorted by purpose. The first, intended for researchers, contains scientific maps accompanied by scientific reports and literature. The second is pedagogical, for promoting the teaching of geography, historical geography and Scriptural Geography at schools and also included wall maps.

Relative to the existing Palestine cartographical works, Kiepert produced the most accurate maps. Unlike Berghaus, the master of contemporary cartographers, Kiepert's skills lay in his knowledge of history and the Orient, in his ability to analyse critically maps and surveys compiled by other scholars and in his ability to learn and apply new methods. This expertise enabled him to draw maps rapidly and ranked him for many years among the most distinguished scholars and cartographers.

By 1841, the twenty-three-year-old Kiepert had completed a one-year expedition to Asia Minor and Greece in order to complete the work begun by Major Helmut von Moltke and his fellow officers, all of them Ritter's students in the *Allgemeine Kriegsschule* who served from 1835 to 1839 as military advisers to the Ottoman army.[181] Kiepert most likely left Berlin only after completing his work for Robinson, as Williams noted that by 1840 *Biblical Researches* was virtually complete. Smith and Hitchcock were more precise, writing that the manuscript 'was ready for the press' in August 1840.[182] Both were correct with regard to the book, but they failed to consider the tremendous difficulties caused by the production of the maps.

Producing the Maps for the *Biblical Researches*: The Robinson-Kiepert Collaboration

It is quite clear that Robinson thought a great deal about the maps from the outset of his work in Berlin. In his detailed letter of 22 January 1839, he consulted with Smith about the geographical areas to be covered by the maps:

> In respect of maps & plans, it will be necessary to cover the whole ground from Mt. Sinai to Beirût; this may probably be done on 2 or perhaps 3 sheets, including Lower Egypt, or the E. part of it. Then there must be a piece of the region around Sinai; of the S. part of the Dead Sea, including the peninsula; of Jerusalem; & perhaps the region around Nâblûs.[183]

One of Robinson's letters to Ritter appears to be a key document for understanding the map-making process, contradicting the accepted version of events. Kept in the latter's papers at the Berlin State Library and dated 2 November 1840, it confirms that the production of the maps by Kiepert had been far from simple and easy. Robinson wrote from London:

> On the eve of embarking for my home across the Atlantic, I would like to inform you of the progress of my maps in Berlin [...]. Of the progress of my Maps in Berlin Mr. Mahlmann will have informed you; & also that not withstanding all our efforts, as well as your own, & not withstanding the most solemn promises, the drawings were *not received* in *time* to complete the engraving before my departure; – & when they did come, they were only piecemeal, & in such a state, that only a skillful cartographer like Mr. Mahlmann could ever have engraved them. This great unfaithfulness on the part of

> Kiepert has been very painful to me, as it shut out the possibility of examining & correcting the maps as I wished; so that I fear many things which he has inserted are untenable![184]

One cannot underestimate the pain and disappointment behind these difficult words. Robinson was once again faced with informing Ritter that his cartographer had let him down. Fortunately, Mahlmann was able to cover and produce the work according to Robinson's wishes and standards.

The two Mahlmann brothers were very close to Ritter. Carl Heinrich Wilhelm was his colleague and expert on mathematical geography, publisher of Humboldt's *Central Asia* (1844) and also involved with various publications of scientific reports on the Holy Land.[185] His brother, Heinrich Christian Ludwig, engraved many of the maps. Something happened in Berlin between 1839, when the 21-year-old student Heinrich Kiepert accepted the work, and the writing of this letter, causing Robinson to write that Mahlmann 'forms a perfect contrast to Kiepert', with his 'skill, fidelity, & punctuality', and that 'I have hardly ever met a man in whom I repose more confidence.'[186]

The same letter is evidence of another interesting fact. This severe criticism did not prevent Robinson and Kiepert from continuing their collaboration. The very same day that Robinson complained to Ritter quite bitterly about his *protégée*, he wrote a long and detailed letter to Kiepert in Berlin. Robinson gave detailed instructions for the continuation of the cartographical work, adding to a list of cartographic instruction also that 'the maps for England will have to be sent over *uncoloured* on account of the double duty, except two or three copies, as models for those to be done by here'. Robinson asked for corrections to certain routes, approved Kiepert's division into districts ('not in Pashaliks which no longer exist under the Egypt[ia]n Gov[ernment]') and added eight specific changes needed of place names, locations, etc.[187]

Robinson was well aware that he needed additional material for the maps and that the expedition data needed confirmation from other sources. In Berlin, he met with Gustav Friedrich Konstantin Parthey, who travelled in Egypt and the Holy Land in 1822–3 with his companions Peter von Medem and astronomer Johann Heinrich Westphal. Westphal's measurements in and around Jerusalem were used for a map of the city published in 1825 in the first volume of *Hertha*, a periodical founded and edited by Berghaus.[188] Parthey, who inherited Berlin's *Nicolaische Buchhandlung*, devoted most of his time to his own scientific research, published extensively in the leading German geographical periodicals. He was an active member of the Geographical Society who Robinson described as 'a man of solid learning, & by inheritance a bookseller'.[189] Parthey gave Robinson 'the angles etc. taken by his companion Westphal, with a sextant'. 'Perhaps something will be made of them,' he added.[190]

Robinson had to deal with a somewhat delicate side issue when Karl von Raumer sought to offer his map of Palestine for use in his work. A natural historian from Munich, Raumer began studying Biblical Geography in 1829 and collaborated with Berghaus, although he never visited Palestine.[191] He published the first edition of his *Palästina* in 1835 and the second edition, much larger and more detailed, in 1838. This edition included maps and a paper discussing the route of the Exodus.[192] Robinson wrote to Smith in January 1839 that Raumer, who worked with the Gotha publishing house of Perthes, had sent his manuscript map of Palestine to Ritter, with a request that Robinson review it. Robinson's reaction was strongly negative. 'In its physical features', he wrote, 'it is copied from Berghaus; in the historical part, he has put in much at random all the names he could gather from the scriptures & others.' Berghaus wrote to Justus Perthes in Gotha 'to tell him that there must be an entirely new map as a result of our research. The matter now rests in Ritter's hands till further wind comes from Perthes.'[193]

After the maps had been compiled and engraved in Berlin, they were presented to various scholars for evaluation. In London, Robinson brought them 'before Col Leake & several of the leading geographers & publishers' and they 'met with very decided approbation'.[194] This enabled the pedantic Robinson to finally give his own approval. In Germany, he advised Kiepert to present them before 'Prof. Ehrenberg, as well as Dr Ritter & Perthes'. The scholars consulted were very distinguished. William Martin Leake, British artillery officer trained in surveying, served in the Near East between 1802 and 1811, primarily in espionage and other secret missions. He retired as a colonel in 1815, made a name for himself through his topographical and antiquarian studies and was a founding member of the RGS.[195] He maintained regular contact with Ritter and Kiepert, including on the latter's mapping of Asia Minor.[196] The physician Friedrich Wilhelm Hemprich and the microbiologist Christian Gottfried Ehrenberg embarked during the 1820s on a six-year research and collection expedition to Egypt, Libya, Sudan, Ethiopia, Arabia, Palestine and Lebanon. They were sent and financed by the Prussian Academy of Sciences, and collected, preserved and shipped to Berlin '114 boxes of 20 to 30 cubic feet' in volume. Ehrenberg published their results, including identification of 110 types of corals, herbarium of 2,900 specimens, 300 rocks and more than 3,500 species of fauna. Ehrenberg became a leading natural historian and a highly distinguished member of the Academy of Science.[197] His maps led to many cartographic renovations of the regions they visited.[198] The other German expert consulted in addition to Ehrenberg was Wilhelm Perthes, son of Justus Perthes, who in 1785 established in Gotha the Geographical Institute publishing house named after him.[199]

Mahlmann's work was so well appreciated that John Murray, Robinson's London publisher, asked the German cartographer to engrave the maps for the English publication as well.[200] Robinson was also worried, however, about the impact on his future collaboration with Kiepert, who might have been insulted by the fact that his work had been given to another professional. Consequently, he added in his letter to Kiepert, dated 2 November 1840, that 'they [the maps] will thus contribute to do great credit to your name & his [Mahlmann] in Germany, England and America'. Robinson promised 22-year-old Kiepert no less than world fame, after showing him that, while he might be talented, he was not talented enough and was very undependable.[201] In contrast, Robinson showed his confidence in Mahlmann and his work, writing to Ritter that 'it will be something quite new for Germany, to have England coming to here for the engraving of maps.' This expressed confidence, will surely 'motivate and stimulate' Mahlmann to 'produce [for the London maps] a work still more beautiful than the German copies'.[202]

Robinson also expressed great appreciation for Murray's skills: 'I must regard it as about to appear before the British public under the best auspices.' Ritter also praised his work and wrote to Robinson that Murray, naturally looking for the best publicity, 'is very desirous to be permitted to make public use of it, with your name.' Robinson asked Ritter to permit both publishers, Murray as well as the Waisenhaus publishing house in Halle, to make use of his recommendation of the maps and under his name. He even dared to ask Ritter to 'send a line' in this respect to Murray, as quickly as possible.[203]

From these letters, one may assume that by late 1840 the work had already been completed. However, Robinson was continually checking and rechecking and consequently revising. He added a note to his 21 December 1840 letter to Ritter, asking him to forward it 'without delay' to Mahlmann, as 'it contains some corrections for the maps'.[204] Kiepert planned to produce a small school map of Palestine for use in Ritter's lectures and Robinson advised him to wait about six months, 'perhaps at Easter', before seeking permission from 'Director Niemeyer' of Halle Waisenhaus Publishing, where the German edition was due to come out.[205] It seems that Kiepert did not accomplish this plan.

Added to the German edition in 1841 were a five-sheet atlas and Kiepert's *Memoir.* Its text is divided according to regions and sub-regions and cites the numerous sources used to compile the map in addition to Robinson's and Smith's itineraries:

Kiepert: Sources of material, connections, data-transfer.

General:
Berghaus, Memoir; Berggren, journal; Bertou, journal; Bird, map and manuscript; Buckingham, journal; Burckhardt, report and journal; Ehrenberg, map and manuscript; Jacotin, survey-map; Laborde, map; Mahlmann, engraver; Moresby, survey-map; Seetzen, map; Prokesch, journal.

For the Peninsula of Mount Sinai and Arabia Petræa:
Berghaus; Bertou; Burckhardt; Irby and Mangles; Jacotin; Laborde; Moore; Moresby-Carless; Prudhoe; Rüppell; Walker.

For Jerusalem and its environs:
Berggren; Catherwood; Doubdan; Pococke; Prokesch; Sieber; Westphal.

For Palestine:
Ancient itineraries; Berggren; Berghaus; Bertou; Buckingham; Burckhardt; Corry; Ehrenberg; Gauttier; Hell; Irby and Mangles; Jacotin; Lanneau; Mariti; Medem and Parthey; Moore; Niebuhr; Paultré; Seetzen; Smith 1834.

For Mount Lebanon:
Barker; Berghaus; Bertou; Bird; Burckhardt; Buckingham; Ehrenberg; Maundrell; Squire.

Kiepert also highlighted the more significant sources, including Berghaus concerning the geographical positions and Jacotin, Ehrenberg, Buckingham, Burckhardt, Irby and Mangles and Schubert. The list includes some interesting names. One of them is James Silk Buckingham, British traveller, author and journalist, who in 1815 embarked on a long expedition in the Near East, in 1818 settled in India and in 1822 published a book about his travels. Smith and Robinson repeatedly criticised and attacked Buckingham's views and geographical notations after his lecture series was published in the *New York Observer*. Smith wrote to the editors that 'the original object of my strictness upon his lectures was to show sensible & especially religious persons that his authority was good for nothing in matters pertaining to Biblical geography & antiquities'.[206]

KIEPERT: THE INTERIM YEARS

Kiepert's involvement in Palestine cartography became increasingly intensive in the years following his initial collaboration with Robinson. He became a leading expert on maps of the region, in its antiquity as well as in the contemporary period and compiled various wall-maps for use in Bible education. His maps, as can be seen in the various bibliographies, accompany publications by many scholars who have studied the Holy Land, Syria, Sinai and Trans-Jordan.[207]

Certainly, the maps produced for the various volumes of *Biblical Researches* were the beginning of this life-long engagement with Holy Land cartography. For example, when producing a second *Map of Old Palestine*, Kiepert noted many names of places and regions that were essential to include and he took them from the second, corrected edition of his *Map of Palestine* compiled for Robinson and Smith (see Plate 4).[208] He continued to improve his maps with each and every piece of data that reached Berlin. To the second edition of the *Map of Old Palestine*, published in 1843 (first edition 1842), he incorporated Major Frederick Holt Robe's pioneering map of the region around the sources of the Jordan. Robe's map was first published in 1842 in the journal of Berlin's Geographical Society through Ritter's mediation and in the following year by Robinson in the *Bibliotheca Sacra*.[209] Robinson continued to improve his own maps and reissue the corrected versions and helped other cartographers publish their maps, all of them based on Kiepert's work.[210]

In 1845, Kiepert accepted the post of head of the Geographical Institute in Weimar, where he stayed until 1852. Although he continued to publish, he considered these years a period of scientific seclusion. An extensive letter exchange, housed in different collections of Berlin State Library, reveals Kiepert's continued efforts to maintain his connections with academic and scientific centres. He discussed maps with Ritter and corrected them, using travellers between their cities to transfer the manuscripts between them.[211] He also maintained his involvement with Palestine research and cartography. An interesting example of these efforts is his meeting with the *'Orientalischen Reisenden'* 'Ingenieur Zimpel'.[212] After the German railroad engineer Charles Franz Zimpel returned home after working for many years in North America, he turned to religion and became one of the strongest Christian advocates of the 'restoration of the Jews'.[213] Zimpel visited Kiepert as early as 1849, probably in preparation for his planned travel. He visited Palestine in spring 1852, and in 1853 published his book attempting to identify sacred sites anew, primarily those connected with Christ's suffering. His map of Jerusalem was 'based on the plan of the English naval lieutenants Aldrich and Symonds, [and] the historian Josephus Flavius and A. C. Emmerich', to which he added corrections he had made 'on the spot'. This was a unique combination even for Palestine literature, which has almost always been affected by geo-religious considerations. Even in this context, the combination of the British naval officers whose map of Jerusalem had recently been published, the historian Josephus Flavius (AD 38–*c.*100) and the ecstatic Anna Katharina Emmerick, 'the nun from Dülmen', whose book served as a geographical guide for Palestine pilgrims, was highly exceptional.[214] Though Zimpel did meet Kiepert, there is no mention of his maps. In the 1860s, Zimpel developed a relationship with August Petermann and Justus Perthes in Gotha, to whom he presented his plan for railways in Palestine using maps drawn by Conrad Schick.[215]

In 1852, Ritter desperately needed a cartographer to collaborate with him on his monumental life project, the *Vergleichende Erdkunde*, particularly on Palestine (published since 1848) and Asia Minor. None of the other Berlin cartographers seemed capable of accomplishing the task and Ritter summoned Kiepert back to Berlin. Ritter was probably also responsible for Kiepert's employment as the house cartographer of Georg Reimers' publishing house. This may have been part of a larger contract, as from then on this firm published most of Kiepert's work. After being, one can imagine, repeatedly disappointed by the quality of work of other cartographers, or by the lack of availability, as in the case of Berghaus, Ritter returned to the only individual who met his standards and with whom he had a successful collaboration. Seven years later, Kiepert was chosen to replace his deceased teacher in the chair of geography at the University of Berlin.[216] His position within the geographical academic milieu was firmly established in 1862, when the RGS accepted him as a corresponding member, as 'he is without doubt the first amongst German geographers', replacing Humboldt and Ritter.[217]

CARTOGRAPHY TO ROBINSON'S SECOND EXPEDITION

Robinson and Smith understood that despite their efforts in gathering considerable data for their maps, significant data was still missing. In order to bridge that gap, they organised measuring instruments for Smith to take with him upon his return to the East from the United States, as promised by Robinson:

> In the preface, written at Berlin, I mentioned that you return to the field of your labors, furnished with instruments for prosecuting further inquiries begun by us; I express the hope of being the medium for communicating those well to the public; that in this way, as we have begun, so we may continue to labour together for the illustration of the Word of God. May this hope be fulfilled![218]

Unfortunately for both, Smith returned to Beirut in April 1841, 'but the state of confusion and anarchy and war since existing in Mount Lebanon [...] has also hitherto cut off all opportunity for travelling and personal observation on his part.'[219] The reference is to the second Syrian campaign, when European military forces, primarily from Britain and the Habsburg Empire, helped the Ottomans to withdraw Muhammed 'Ali's Egyptian forces back into Egypt.[220] Naturally, many of the letters, primarily between the missionaries, deal with the war and its effects.

In the end, Smith was able to use his instruments and supplied Robinson with material for more maps, which the latter called 'your maps'. The data was used by both Berghaus and Kiepert for various projects, many concerning Lebanon. They included the routes on a scale of 1:100,000 or greater and a map on a scale of 1:200,000, reduced for engraving to 1:400,000. Robinson left much discretion to 'our engraver, Mahlmann', who is 'very intelligent, & himself a scientific geographer'.[221] Kiepert's first map containing Lebanon was published only ten years later for Robinson's *Later Biblical Researches*.

Robinson suggested compiling maps for popular use based upon the existing ones, hoping they would also sell better. In 1846, he proposed to Smith 'to publish a cheap general map of Palestine, ancient & modern, based mainly on Kiepert's plan, but larger, say 1:400,000, as in the Researches, or about 3 ft high by 2 or 1⅓ ft wide [...].' Engraved on stone and printed using a lithographic process, it could be multiplied indefinitely at a low cost. Always keeping the missionary purpose in mind, Robinson suggested that 'a cheap map thus brought out under your supervision & with your name, would rapidly sell by thousands, & do more to diffuse a current knowledge of the country, than anything else.'[222]

Robinson was very involved in the problem of the Jordan Rift Valley. He and his colleagues followed the various projects initiated to solve the still unanswered questions:

> I may say, perhaps *to you*, that our Expedition, however well they have done so far as they meant, have yet *settled* nothing, as to the depression of Jordan & its valley, except

> possibly the depression of the Dead Sea. They ought to have leveled across to the Lake of Tiberias, & thence along the Jordan to the Dead Sea [...].[223]

Consequently, they focused on this issue in their publications and in their correspondence. Robinson's important role in the realisation of William Francis Lynch's American expedition to the Jordan and the Dead Sea in 1848 ('To J.[sic] Robinson [...] I was indebted for letters to his friends in Beirût, and for much information furnished from his copious store') and the public discussion of its justification is described in detail in the letters and still awaits the much needed paper.[224]

Robinson never gave up the idea of another expedition to the East. However, it took some years before he began mentioning it in his letters. After discussing, as usual, all the new studies and books concerning the Holy Land, those that he received and those of which he only heard about and reviewing the various questions he was researching, he concluded his letter of 31 March 1847 with the hope of taking a year of leave, most likely in 1849:

> All these things only increase my longings once more to revisit the East, & carry out at the spot these inquiries with so much mixed *sober* observation. So strong is my desire that not the *will*, but the liberty alone is wanting this very year. But I say: 'If the Lord will, I shall live, & do this, & thank', perhaps, another year [see Fig. 1.5].[225]

On 20 August 1851, Robinson was in London after a family visit to Germany. It took him a couple of weeks to make up his mind and draw up the plan; on 1 March 1852, he planned to arrive in Beirut and spend the summer in the Holy Land. 'You see,' he wrote to Smith, 'my desires and plans for again visiting Palestine, have not been all talk; there has always been an earnest purpose of doing so, if God in his providence should seem to open the door.'[226] Robinson understood that Smith was unable to accompany him for the entire time: 'I know well the absorbing nature of your work of translation; as well as its unsurpassed importance; if I could but aid you in it in any way I should regain it a high privilege.'[227] Robinson's letters are filled with detailed ideas and preparations for the expedition, another topic that is beyond the scope of this chapter. On 24 January 1852 he wrote from Berlin and on 3 June he was in Damascus, having brought with him as printed source-books 'a copy of Ritter, the new edition of Raumer, perh[aps] a copy of Reland, & of Irby & Mangles'.[228] The short trip ended three weeks later in Smyrna and Robinson wrote, 'thus has ended my (or rather *our*) second tour in Palestine! I look back upon it almost as a dream.' On his return trip, he planned 'to sit down with Kiepert for a time'.[229]

Whereas in 1840 Kiepert was an extremely young and inexperienced cartographer, twelve years later, when Robinson returned from his second expedition to the East, the situation was totally different. The existing letters reveal that Robinson had no second thoughts and from the beginning was planning on collaborating with Kiepert. Halle, where

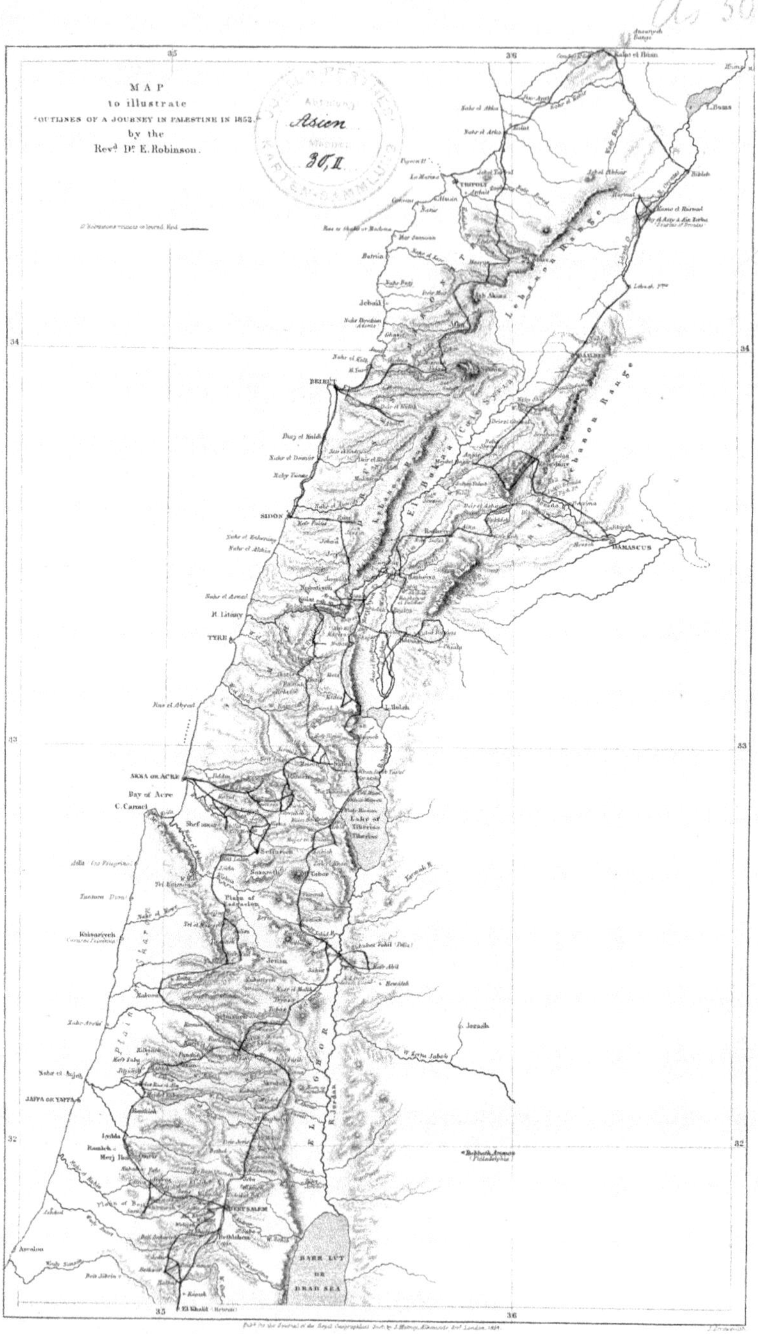

Figure 1.5 John Arrowsmith, *Map to illustrate 'Outlines of a Journey in Palestine in 1852' by the Rev. Dr. E. Robinson* (London, 1854)
Source: FBG SPK 547-112619673

Robinson visited his family and friends, was not that far from Kiepert's 'exile' in Weimar, so Robinson simply brought him 'a package of Dr Smith's Notes on a Tour in Northern Syria (7 small sheets)'. Two more packages 'completing the same in 20 sheets in all' were sent to Ritter by post, 'as you can at present make more & better use of them than anyone else'. Robinson expected Ritter and Kiepert to work together. Robinson claimed that Kiepert had all the necessary material and expected him to cooperate with 'the younger Mr Reimer'.[230] It was perfect timing, since Ritter and Kiepert were expected, precisely in the winter of 1852–3, to work on the maps accompanying the latter's 'volume for Northern Syria & Phoenicia', and it was 'precisely where these notes & Itineraries belong'. Accordingly, 'in Weimar my time was wholly occupied in the preparations of the material for the map'.[231]

And what a wealth of material! Robinson had recruited all missionaries, American and others, to help him collect as much material as possible. They did respond, some of them travelling with him, some taking on specific missions, others taking notes during their travels. There were missionaries whose reports were highly dependable and consequently of great value. 'The papers are a proof of a diligent observer,' wrote critical Tobler about Wolcott.[232] Josias Leslie Porter, an Irish Presbyterian Church missionary stationed in Damascus for ten years from December 1849, contributed significantly to the study of the regions around Damascus, the Golan and Hauran.[233] Following Robinson's request, he delivered Smith 'a list of names of the villages in the plain of Damascus [...] the names arranged under the heading of the districts in which they are situated,' and added a 'list of the villages in the plain of Damascus, Jaulain and Jeiduns'. He also offered to add the number of inhabitants in each village.[234]

This is but one example. There is no better testimony than Robinson's own views, from another letter of his to Smith:

> Kiepert [...] has completed the general construction of the map; using at the same time the poor works of Schultz; & the far more valuable one of Callier, the French Engineer who accompanied Michaud & Poujoulât in their oriental journey; this, although the journey was made 20 years ago has never been published; it has been engraved, & Ritter has a copy. [...] Ritter has also the great net of Symonds' triangulation; & this too Kiepert has used.[235]

Robinson noted Schultz's 'poor works', as Schultz worked together with Williams, Robinson's greatest antagonist (see Fig. 1.6).[236] Camille-Antoine Callier was an army officer and cartographer sent to the Near East by the French War Ministry. He travelled with Joseph-François Michaud, poet and historian of the Crusader period and member of the French Academy and his companion Jean Joseph François Poujoulat.[237] The triangulation was executed by the young British Royal Engineer Lieutenant John Frederick Anthony Symonds in 1841–2 according to two baselines, the first from Haifa to the Sea of Galilee and the second from Jaffa through Jerusalem to the Dead Sea. The sketch reached Germany

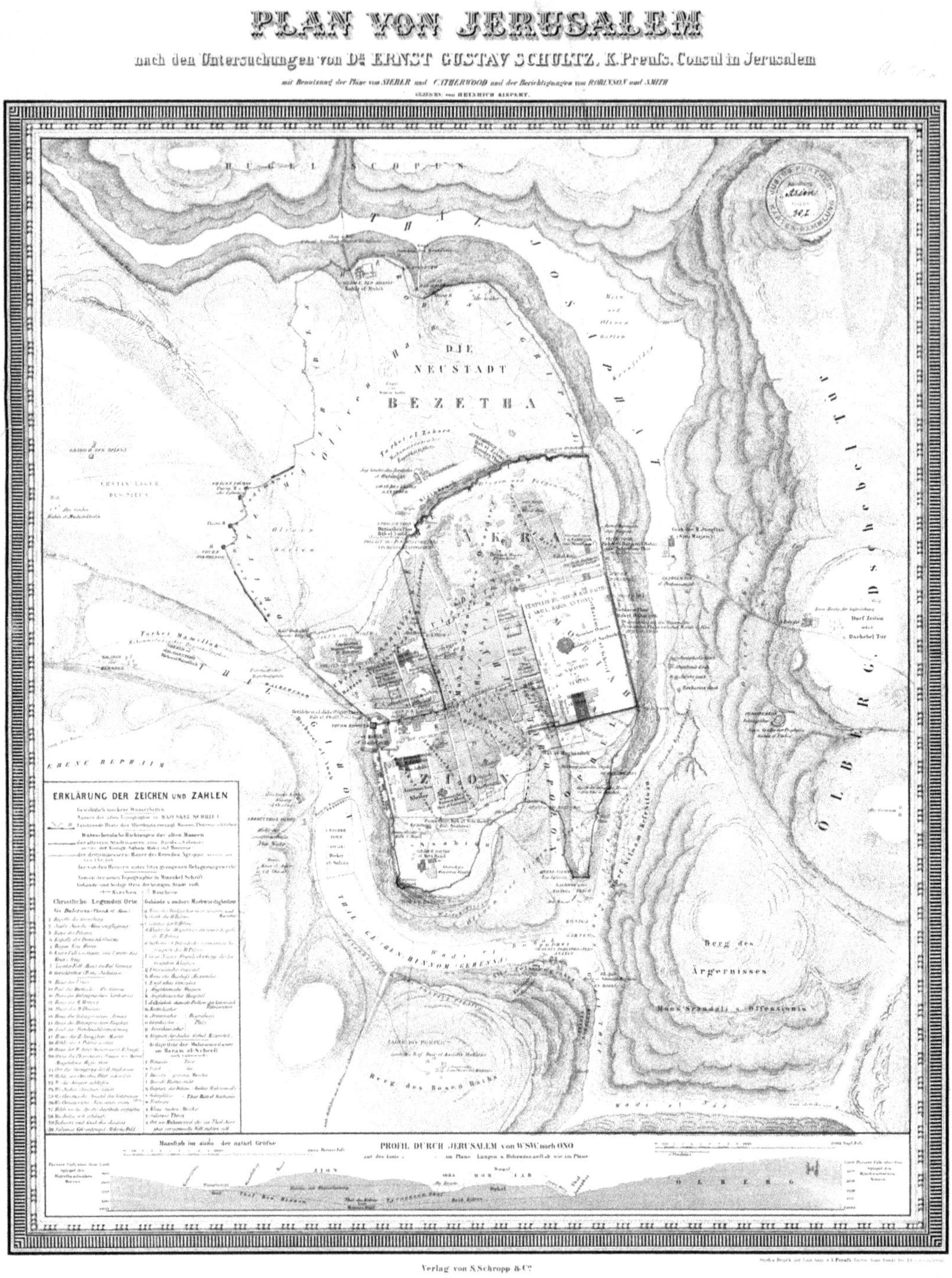

Figure 1.6 Heinrich Kiepert, *Plan von Jerusalem nach den Untersuchungen von Dr. Ernst Gustav Schultz, K. Preuss. Consul in Jerusalem ...* (*Plan of Jerusalem according to Ernst Gustav Schultz*) (Berlin, 1845)
Source: FBG SPK 547-112557015

through the mediation of the Dutch cartographer Charles William Meredith van de Velde, who will be dealt in detail by Jutta Faehndrich in the Chapter 2.[238] However, the Dutch cartographer, who brought Ritter's relevant volumes and Robinson's *Biblical Researches* with him to Palestine, as both were 'indispensable', and who, as shown in various letters, had been helped by Ritter, Robinson and Smith, mentions the maps in his *Memoir* only coincidently and briefly. The fact that he discusses other maps at length raises a rather large question concerning their relations.

Despite the wealth of data that Robinson collected, Kiepert had complaints concerning the exactness of routes recorded in most of the itineraries, as well as the absence of solid itineraries for some of the special, central roads such as Beirût–Damascus or 'Akka–Tyre. Therefore, he cautioned that missionaries must remember 'that every step they take *may* throw more light upon the cartography of Palestine; & that they will always take notes exact enough to work out from them a map of the route'.[239] The letters also include much information concerning van de Velde's obsessive travel throughout Europe, so well described by Jutta Faehndrich (see Chapter 2), looking for every piece of information that could help him with the publication of his map.[240] According to Robinson, van de Velde approached Ritter, stating that he had access to Symonds' maps, which he was tracing; his main objective probably 'was to draw out an invitation to come to Berlin'. Van de Velde asked Ritter's advice for the best place to publish his map, which he did in Gotha by Justus Perthes.[241]

As mentioned above, scholars who planned research expedition in Asia Minor and Palestine, arrived at Berlin to consult Kiepert and get cartographic instructions. In this respect, too, he was following Ritter, who, as also mentioned above, was a centre of data, guidance and publication for Holy Land explorers. Kiepert's instructions usually included maps and sketches depicting the regions he considered to be less explored, those from which he needed more information, and detailed instructions for field study.[242] Kiepert used in his maps the most up-to-date scientific instruments and he also profoundly analysed and constantly and deliberately compared Robinson's results with data supplied by other scholars or travellers. In some cases he did not hesitate even to criticise his own tutor, Berghaus; hence, at least relatively to the existing Palestine cartographical works, he produced the most accurate maps.

The work on the maps for the 1856 edition should have been relatively easier and calmer. Both Kiepert and Robinson were already highly experienced and also sixteen years older. The letters present a different picture. One good example is this detailed letter from Murray, then in Edinburgh, to Robinson, dated March of that year:

> I am glad to find that you are so far advanced with the printing of the new edition of your travels in Palestine. I shall be glad to receive the stereotype plates. When ready & will defray the cost as agreed on. Meanwhile I have been some time occupied with the maps – & have been in correspondence with Mr Kiepert – from whom I received a few

> days ago the drawing of Sinai & the Delta of Egypt. This I have brought down with me to this place in order to put it into the hands of that excellent geographer Mr Keith Johnston to engrave[?].[243]

Alexander Keith Johnston, Scottish geographer and cartographer, founded in 1826, together with his brother William, the cartographical firm W. & A.K. Johnston. He had longstanding connections with German cartographers, beginning with Berghaus.[244] The question is why Murray took Kiepert's drawings to Edinburgh. The clear answer lies in the following sentences of his letter:

> I am sorry to say he has found Kiepert's drawing very defaced(?). He seems to have been ignorant certainly to have made use of the recent French survey – & is very faulty in many of his positions – as I have been myself convinced by the comparison.[245]

The recent French survey that Murray alluded to in 1856 was mentioned by Robinson to Smith in a letter dated 20 August 1853:

> Have you seen the book of the Frenchman De Saulcy, who found the five cities of the place of Sodom? It is the most perfectly French book I have seen for many a day, full of argument & illusions, enormous blunders, & many evil things; & yet I fear some will credit his fables.[246]

The French officer and scholar Baron Louis Félicien Joseph Caignart de Saulcy travelled in the Holy Land accompanied by scholars in 1850–1. His book included various maps, measured and drawn by cartographer Charles Gélis, an army officer who performed all the measurements during the expedition.[247] But, like many of his contemporaries, Robinson did not take de Saulcy's publications, theories or locations too seriously.

Murray's letter noted additional difficulties. Paul-Adrien Bourdaloue was a French civil engineer who started his career planning railway lines and between 1857 and 1863 he planned and built France's first levelled lines across the country. At the request of the French chief engineer of Egyptian public works, Louis Adolphe Linant de Bellefonds, chief engineer of planning and constructing the Suez Canal, Bourdaloue executed a survey of the region in 1847. His primary objective was to test the supposition of Napoleon's engineers that there is a significant difference in the level between the Red Sea and the Mediterranean. His survey was published in France in 1856 and probably reached London quickly.[248] This supposition also fits the next sentences in the letter, where Murray wrote that he asked Johnston, KJ as he calls him, 'to make a fresh drawing of the Egyptian position as far as Suez'. However, the next sentence does cause a problem, as Murray writes that the 'work place is also more accurately fixed by the E.J. Courpre [reading problematic] survey – than by Kiepert-Johnston charts[?]'. We were unable to locate such a survey of the 1850s. As for the Sinai Peninsula, Kiepert could have improved his map according to John Wilson's *Map of the Holy Land* that he compiled together with a map of Sinai and Arabia Petraea for his 1847 published book *Lands of the Bible*. Both maps were engraved by Johnston.[249]

Murray continued with more technical matters regarding the costs of the engraving and printing:

> I propose to defray also the cost of engraving them in the best manner – then to make transfer on two or more stones – for the American & Berlin Editions [?] either sending over the stones or printing the requested number of copies as maybe deemed most expedient as – I should charge a proportion of the cost of drawing & engraving to the [?] might be considerately reduced if I were allowed the right of employing the map for other works such as Handbook of Palestine. This better suggestion I throw are for your consideration in case you should have objection to it.[250]

In July, when informing Robinson that the books were ready, Murray was still reporting problems with the maps:

> I am very sorry there should be any misunderstanding about the Maps. I had concluded that you had arranged before hand with Kiepert to re-draw them. [...] I hope mine will be better than yours, but will beg you to send me proofs of yours when done as I will to you of mine [...]. I have not yet got the drawing of Palestine.[251]

Nonetheless, the process was quick. On 19 November of that same year, six months after Murray's letter, the RGS secretary could thank Robinson for donating his books to its library.[252]

SUMMARY: BERGHAUS AND KIEPERT MAPS FOR ROBINSON AND SMITH PUBLICATIONS

There are two primary ways to summarise this rather long chapter. The first is to check the reception and evaluation of the maps by contemporary scholars, mainly experts for Holy Land and sacred literature, as well as cartographers who deal with the region. In addition, there is the issue of the extent to which the maps were used for future cartography. Both have been dealt with in former studies. The second mode of summary is to focus on the revisions to the accepted narrative that resulted from this very extensive archival study. The extensive study performed in the German, American and British archives, the thorough and careful reading of hundreds of documents and the piecing together of the documents in order to establish the most authentic narrative behind the process of manufacturing the maps, led to numerous changes in the accepted narrative to date of this crucial stage in Holy Land cartography.

Ritter's Berlin developed into a centre for Palestine studies, probably the largest and best informed in Europe. Naturally, this data, knowledge and material, in whole or in part, were available to scholars interested in Holy Land studies and mapping. Robinson 'discovered' Ritter only relatively late in his studies in Germany, a fact first established by Williams after carefully reading Robinson's letters. This was the first of a number of changes to the accepted narrative concerning Robinson and Smith, their travels to the Holy Land and publication of their books. The numerous handwritten documents, primarily letters, supplied interesting answers to questions concerning these issues and led to changes in the accepted narrative. The examples discussed above are abundant:

- we discovered what seems to be the real reason behind Berghaus's refusal to continue working with Robinson;
- we described the difficulties faced by Robinson in his work with Kiepert on the first book as well as on the second;
- we established Mahlmann's important role in the production of the maps;
- we had a partial glimpse into the huge number of sources used by Robinson and by Kiepert;
- we established the important and even crucial part played by Ritter and his role as teacher and mediator;
- we pointed at some aspects of mid-nineteenth-century international scientific connections and work as well as at some rivalries; and
- we received new insights into some of the most important figures within European Holy Land and Sacred Literature scholars.

Unlike Berghaus, the master of contemporary cartographers, Kiepert's skills lay in his historical knowledge, oriental studies and his ability to analyse and criticise maps and surveys performed by other scholars and in his ability to study and adapt new methods. These elements allowed him to draw maps rapidly and ranked him among the most distinguished scholars and cartographers for many years. His involvement in Holy Land cartography and in Biblical archaeology started, as demonstrated in this chapter, in his projects for and with Robinson. In the years that followed, his involvement became increasingly intensive and led to his recognition as an expert on maps of the region, of its antiquity as well as contemporary times. His maps, as can be seen in the various bibliographies, accompany publications by many scholars who studied the Holy Land, Syria, Sinai and Trans-Jordan.

The maps were copied, used and improved by many scholars, travellers and cartographers and this is another issue for further study. As examples, cartographer Carl Helmuth from Halle published as early as 1843 two maps, of the Holy Land and of Jerusalem, both mainly 'after' Robinson and Smith.[253] There were additional maps, and not necessarily German, such as Joseph Tracy's map of 1845.[254] The German botanist and cartographer Karl Friedrich Robert Schneider was one of many people to use Kiepert's large wall-maps for producing his wall-map for Biblical history.[255] In his *Text zu Land*, Markus Kirchhoff discusses at length what he calls 'a cartographic conversion', the use of many German-made maps by the young Zionist movement; also here, Kiepert's works played a significant role.[256] In his recently published book depicting *Hebrew Maps of the Land of Israel*, Rehav (Buni) Rubin shows how deep and extensive was the use of Kiepert's maps by Hebrew cartographers and map copiers from the mid-nineteenth century.[257]

'Already the maps for the first travel-books, 1840,' wrote Tobler in his epoch-making *Bibliographia*, 'surpassed through their correctness and accuracy everything that came out in the field of Palestine cartography. [...] The names of the first-rate observers and the excellent cartographer vouched for the fact, that a very significant progress can be registered.'[258] Being always strict, clear and highly critical, one should not underestimate his compliments. Some years earlier, in the *Memoir* accompanying his and van de Velde's plan of Jerusalem, Tobler had not been so enthusiastic, writing that 'the new materials collected [by Robinson and Smith in Jerusalem] were not indeed considerable', and other sources should be consulted.[259] Ritter's praises have been brought in different places through this chapter and his words were always full of compliments; after all, it was the team that he organised. The task to produce a new map of Palestine, he wrote in his review to Palestine maps, 'was accomplished by H. Kiepert in so masterly a manner and in every respect so thoroughly scientific a spirit, as to win the applause of all scientific judges and to be the model for all following works of its kind'.[260]

These sentences of Ritter, translated into English by his Scottish student William Leonard Gage, form a perfect ending for this chapter.

APPENDIX: BIBLIOGRAPHY OF BERGHAUS'S AND KIEPERT'S MAPS TO ROBINSON AND SMITH

A central goal of our project is to compile a new list of the maps of the Holy Land and/or parts of it, with its immediate environs, based on the above-mentioned bibliographies, as well as other lists and mainly on work in different map collections. Presented here is the existing list up until Robinson's death, the maps created for the *Biblical Researches* and following their publication. Williams completed the tremendous task of compiling a bibliography of 'The Works of Edward Robinson', but with no mention of the maps.[261]

Berghaus

1 Part of Arabia Petræa and Palestine By Heinrich Berghaus 1839, *JRGS* 9 (1839)

Sketch maps

2 Karte von Palæstina vorzüglich nach den Itinerarien von E. Robinson und E. Smith, 1838, mit Benutzung der Jacotin'schen Aufnahme, und der Itinerarien von Burckhardt, Buckingham, Irby & Mangels, von Prokesch, Berggren, E. Smith (1834), Bertou u. a. so wie der Seetzen'schen Karte, construirt und gez. von H. Kiepert ([Berlin], 1840), hand-drawn and coloured, 1:200,000, 82 × 146 cm. [SBB-PK Kart. D 6940] (see Plate 2)

3 Edward Robinson's Routen im Libanon 1838. [1840]. Gez. von Heinrich Kiepert, hand-drawn and coloured sketch, 42 × 61 cm (no scale). [SBB-PK Kart. D 7412], war loss [Kriegsverlust]

Palästina und die südlich angrenzenden Länder (Halle, 1841)

(equally issued in *Atlas zu Robinson's Palästina*)

4 Karte von Palaestina. Vorzüglich nach den Itinerarien und Messungen von E. Robinson und E. Smith, konstruirt und gezeichnet von H. Kiepert. H. Mahlmann sc. (Berlin, 1840). 1:400,000; 37 × 84 cm (two sheets, each 37 × 42 cm) (see Plate 3)

Insets:

- Der nördliche Theil des Libanon, nach Ehrenberg und Bird, 1:400,000, 17.9 × 14 cm
- Umgebungen von Beirût, 1:200,000, 7.5 × 11.7 cm
- (See Genezareth oder Tiberias),[262] 1:200,000, 14.2 × 9.9 cm

5 Karte der Sinai-Halbinsel und des Petraeischen Arabiens. Nach den Itinerarien von E. Robinson und E. Smith konstruirt und gezeichnet von H. Kiepert. H. Mahlmann sc. (Berlin, 1840). 1:800,000. 42 × 54 cm

Insets:

- Die Gegend zwischen Kairo und Suez, 19.5 × 17 cm
- Pass von 'Akabah. 1:200,000, 6.2 × 8.1 cm
- Plan von Petra nach Laborde. 1:20,000, 9.1 × 9.8 cm

6 Der Sinai, H. Kiepert del., H. Mahlmann sc., 1:100,000. 13 × 18 cm

7 Plan von Jerusalem, entw. nach Sieber und Catherwood, berichtigt durch Beobachtungen von Robinson und Smith. Entworfen u. gezeichnet v. H. Kiepert. In Stein gest. von H. Mahlmann; 1:10,000, 25 × 30 cm

8 Umgebungen von Jerusalem, nach Messungen von Robinson und Smith konstruirt von H. Kiepert. In Stein gest. von H. Mahlmann; 1:100,000; 15 cm × 30.5 cm[263]

Biblical Researches (London, 1841) (Boston, 1841)[264]

9 Map of Palestine. Chiefly from the Itineraries and Measurements of E. Robinson and E. Smith. Constructed and Drawn by H. Kiepert. Engr. on Stone by H. Mahlmann (Berlin, 1840), 1:400,000, 2 sheets, each 37 cm × 42 cm

Insets:
- Northern Part of Mount Lebanon, from the Maps of Ehrenberg and Bird, 1:400,000
- Environs of Beirût, 1:200,000
- Lake of Tiberias and its Environs, 1:200,000

10 Map of the Peninsula of Mount Sinai and Arabia Petraea from the Itineraries of E. Robinson and E. Smith. Constructed and Drawn by H. Kiepert. Engraved on Stone by H. Mahlmann. Berlin 1841, 1:800,000, 41 × 53 cm

11 Sinai, H. Kiepert del. (1841), 1:100,000, 12 × 18 cm

12 Plan of Jerusalem, sketched from Sieber and Catherwood, corrected by the Measurements of Robinson and Smith. Drawn by H. Kiepert. Engraved on Stone by H. Mahlmann. Berlin 1841, 1:10,000, 25 × 30.5 cm (see Fig. 1.4)

13 Environs of Jerusalem, from the Routes and Observations of Robinson and Smith. Drawn by H. Kiepert. Engraved on Stone by H. Mahlmann. Berlin 1841, 1:100,000, 15 × 30.5 cm[265]

Later Biblical Researches (London, 1856) (Boston, 1856)[266]

14 A new map of Palestine including also Phœnicia and Cœlesyria, to accompany Prof. E. Robinson's Later Bibl. Researches constructed principally from unpublished materials furnished by Revd. Ed. Robinson and Revd. Eli Smith, and other American travellers; by Henry Kiepert. Engraved by W. & A.K. Johnston. Edinburgh (London, 1856), 2 sheets. 1:600,000, 37 × 70 cm

Insets:
- Sketch map of Jerusalem reduced from the plans of Symonds & Tobler with many corrections, 1:15,000
- Environs of Jerusalem. 1:200,000

Neuere Biblische Forschungen (Berlin, 1857)

15 Übersichts-Karte von Palästina und Phönicien, vorzüglich nach den Messungen und Beobachtungen von Ed. Robinson und Ely Smith, und mit Benutzung aller bis jetzt

erschienen Reisewerke neu construirt von H. Kiepert zur Erläuterung von Prof. Ed. Robinson's zweiter Reise in Palästina. 1:800,000. 28 × 52 cm

Inset:

- Umgebungen von Jerusalem, 1:200,000

Reprints and derivatives

16 Karte von Palaestina. Nach den neuesten Quellen, vorzüglich nach den Robinsonschen Untersuchungen bearbeitet und gezeichnet von H. Kiepert. Herausgegeben von Prof. Dr. Carl Ritter. Geo-lithogr. Anst. v. H. Mahlmann (Berlin, 1842), 1:600,000, 41 × 56 cm (new improved issue: 1843) (see Plate 4)

Insets:

- Die Sinai-Halbinsel oder Arabia Petraea, 1:2,400,000
- Das Sinai- oder Horeb-Gebirge, 1:200,000
- Alt-Jerusalem

17 Plan von Jerusalem nach den Untersuchungen von Dr. Ernst Gustav Schultz, K. Preuss. Consul in Jerusalem mit Benutzung der Pläne von Sieber und Catherwood und der Berichtigungen von Robinson und Smith. Gezeichn. von Heinrich Kiepert. Stich u. Druck der Lith. Anst. v. L. Preuß Berlin. (gedr. v. E. Berndt) (Berlin, 1845), 1:6.000, 46 × 59 cm (see Fig. 1.6)

Four maps to accompany the Biblical researches in Palestine of E. Robinson and E. Smith (Boston, 1856)

18 Map of Northern Palestine and Lebanon chiefly from the routes of E. Robinson & others drawn by H. Kiepert. Engr. by Lipmann, Brooklyn, N.Y. (Boston, 1856), 1:400,000

19 Map of Southern Palestine chiefly from the routes of E. Robinson & others drawn by H. Kiepert. Engr. by Lipmann, N.Y. (Boston, 1856), 1:400,000

20 Map of the Peninsula of Mount Sinai and Arabia Petraea from the itineraries of E. Robinson and E. Smith drawn by H. Kiepert. Engr. by L. Lipmann (Boston, [1856])

Insets:

- Region between Cairo and Suez
- Plan of Petra from Laborde, 1:20,000

21 Plan of Jerusalem, originally sketched from Sieber and Catherwood, corrected by the Measurements of Robinson and Smith. Engr. by L. Lipmann [Boston, 1856], 1:10,000

22 Environs of Jerusalem, from the Routes and Observations of Robinson and Smith. Engr. by L. Lipmann [Boston, 1856], 1:100,000

23 Karte der Sinai Halbinsel nach den Itinerarien von Burckhardt, Rüppell, Laborde, Callier, Koller, Russegger, Robinson, Abeken und R. Lepsius construirt und gezeichnet von H. Kiepert (Berlin, 1859), 1:500,000, 65 × 51 cm

2 The Land and the Map: C.W.M. van de Velde's *Map of the Holy Land*

Jutta Faehndrich

THE HOLY LAND AND CARTOGRAPHY

The nineteenth century was the golden age of explorers and expeditions. However, the lust for discovery was not about information as such – indigenous people probably knew their surroundings well enough – but helped to construct a decidedly Western body of knowledge about the world. Unlike Africa or the Americas, the region referred to as the Holy Land had been well known since antiquity with its Biblical topography firmly in Catholic and Christian Orthodox hands. Yet the relatively young Protestantism was able to gain ground over the course of the nineteenth century with the help of a Biblical Geography that, allied with historical-critical theology and philology, advanced by establishing knowledge according to newly defined scientific standards.[1]

Nineteenth-century Biblical Geography represented a paradigm shift: in lieu of the unquestioned legendary Catholic topography, it aimed at thoroughly researching geography based on scientific facts. What was more, Palestine was actually a new region for Protestantism and, with the endeavours of scholars such as Edward Robinson,[2] Protestants first and foremost discovered, described and constructed the Holy Land for themselves.[3] This scholarly Palestine was an invention of tradition in the Hobsbawmian sense – a newly created, but not necessarily wrong, 'discovery'. Like other invented traditions of the nineteenth century, it very quickly became popular and ultimately created a collective memory of the region.[4] One of its most paradigmatic works was William Thomson's *The Land and the Book* (1859).[5] The title expresses the idea that the Bible (the Book) and the land together formed an integrated whole – the land could not be understood without the Bible and the Bible could not be understood without the land.[6] This made the geography of the Holy Land quintessential for understanding the scriptures. The title of this chapter alludes to that concept, suggesting that the land and the map together made for understanding the Book.

The history of the Holy Land was shaped by powerful and very effective local attributions of meaning. Indeed, the region is a prime example of how topographies were constructed and symbolic meanings assigned to place and space.[7] These attributions do not date back to biblical times but mostly came about in the nineteenth century when Biblical Geography was being established as a scholarly discipline. Within this process of assigning meaning to place and space, the map as a repository of knowledge played a key role, as did surveying and mapping as inscription techniques. Cartography was – and still is – part of an entire set of practices that describe, inventory and take possession of a region by scientific means. The images constructed this way by far transcended the scholarly discourse of its time, leaving a strong impression on the Western image of the Holy Land even today.[8]

Maps were the most prolific medium of the nineteenth century, perfectly suited to portray the rapid pace of European expansion, industrialisation and modernisation. Journals and periodicals, which could be updated more frequently than the old encyclopaedia, were becoming the media of the latest general knowledge in the nineteenth century.[9] In due course, the map, which was easily revised by simply modifying the copperplate it was printed from, took off on its course as a modern mode of visualising this information.[10] In the case of the Holy Land, mapping the biblical lands was also a prerequisite for the work of archaeologists and other scholars.[11] Together geography, cartography and archaeology were supposed to provide evidence for the scriptures and solve the imminent contradiction between religion and the new era of science.[12] The modern Palestine maps of the nineteenth century are an archive in cartographic form, with, for example, as many contemporary Arabic place names as possible recorded and mapped, which were of prime importance for locating biblical sites.[13] Edward Robinson and Eli Smith, whose pioneering work is so aptly retraced by Haim Goren (see Chapter 1 in this volume), had in fact discovered that Arabic place names often preserved the biblical names due to the similarity of pre-classical Semitic languages to Arabic. When the British Palestine Exploration Fund made a complete survey of the region in the 1870s, collecting indigenous toponyms was one of their most important tasks.[14]

Over the nineteenth century, cartography gradually evolved into a professional trade with its own institutions, rules and regulations. Nevertheless we still know little about the day-to-day workings of nineteenth-century cartographers, especially in the earlier decades. This chapter will also attempt to provide a deeper insight into the incentives, mindsets and predicaments of nineteenth-century cartography by closely studying the creation of a single map.

The Holy Land was a relatively late region to be mapped according to modern standards. Since the end of the crusades, Palestine had been on the outskirts of the Ottoman Empire and far removed from European Christian influence. Thus it had escaped the practice of measuring and mapping that was a characteristic of the modern state.[15] Earlier modern attempts at measuring as much of the biblical lands between Dan and Beersheba as possible were either only incidental, such as during military excursions like Napoleon's, or they were so dangerous that, even after years of preparing themselves for their journeys, many explorers paid with their lives, such as Ulrich Jasper Seetzen and Johann Ludwig Burckhardt.[16] A systematic survey of the region, like the British Empire had carried out in India from 1800 to 1843, remained unthinkable.

In the 1830s, the region started to stabilise politically and economically, not least due to the Tanzimat reforms.[17] Gradually, the safety of European travellers improved. This resulted in explorers, theologians, pilgrims and other travellers becoming more actively involved in the region, for example by measuring and mapping. The *Map of the Holy Land* is part of this Western rediscovery of Palestine in the nineteenth century.[18] After a few respectable earlier maps, by Napoleon's cartographer Pierre Jacotin and by the German Heinrich Berghaus,[19] it presented an attempt at systematically surveying the region. In this respect, the map was in direct competition with the Palestine map that Heinrich Kiepert created based on the measurements of Edward Robinson and Eli Smith described in detail by Haim Goren (see Chapter 1). The cartographers and their publishers were well aware of this competitiveness, as we will see.

THE MAP OF THE HOLY LAND AND ITS CARTOGRAPHER

The map in the focus of this chapter was published in Germany in 1858 and is special in many respects, not least in the extent of correspondence that has survived from its production. The archives of the Perthes Collection, today part of the Gotha Research Library, hold the surviving documents of the Geographische Verlagsanstalt Justus Perthes, one of the most eminent map publishers of its time on the European continent and publisher of our *Map of the Holy Land*. The sources of the Perthes Collection offer a unique glimpse into making a map from its inception, drawing, engraving, proofing, colouring, printing and sale, to a revised second edition and finally to the point when the publisher phased it out of its catalogue twenty years after its initial publishing. The respective correspondence is spread over three bulky folders, holding 426 sheets of letters (dated 1854–66), printing samples and proofs. Almost all of the cartographer's and most of the publisher's letters have survived, the latter in a hard-to-read form of primitive copies made by pressing a transparent sheet of paper onto the letters while the ink was still wet (the so-called *Abklatsch*). We also have the publisher's correspondence with engravers and lithographers as well as with other business partners, and the account books which list expenses and sales. Unfortunately there are no surviving map sketches to be found, only the finished map as it was being processed into further cartographic products of the company.[20]

Since geography and cartography were not national endeavours in the middle of the nineteenth century, nor were the emerging fields of Palestine research and Biblical Geography, the English-language *Map of the Holy Land* could be made by a Dutch cartographer with a German publishing house without any national conflict.[21] In fact, the origins of this project date back to London, which was at that time not only the capital of the British Empire but a bastion of geographical knowledge. In the 1850s, Britain's lively geographic circles, namely the Royal Geographical Society (RGS), brought together two men in their thirties who were both genuinely interested in geography and cartography: August Petermann (portrayed in depth by Bruno Schelhaas in Chapter 3 of this volume) and the Dutchman Charles William Meredith van de Velde.[22] Petermann's Dutch counterpart was a former lieutenant and cartographer who hailed from the Frisian town of Leeuwarden. Born into a navy family in 1818 (see Fig. 2.1),[23] he had joined the Dutch Royal Navy at the young age of fourteen. He there received a thorough training in surveying and cartography but also showed considerable artistic talent. After a career in the Dutch naval surveying department and deployment in the colonies, the Dutch East Indies, he quit service at the end of 1846 as a second lieutenant to dedicate his time to map-making and landscape painting.[24] After his military career, the Frisian was active in the Protestant mission in Cape Colony for two years.[25] From October 1851 to June 1852 he travelled to the Holy Land, back then also known as southern Syria, to make a measurement-based map of the region that had long been considered a desideratum. The resulting *Map of the Holy Land* was published in 1858.[26]

The Dutchman was a fairly colourful contributor to fledgling Palestine research. Like many in his field, he had had military training, not academic. Yet he was clearly driven more by faith than by military ambitions: 'The study of the Holy Scriptures had made me deeply feel the want of a correct and sufficiently detailed map of the land' is how he described his motivation.[27] This combination of faith and military background would later be typical of the British Palestine Exploration Fund (PEF).[28] Van de Velde was also active in Biblical Geography and published a two-volume travelogue in which he identified a considerable amount of biblical sites based on his surveying trip.[29] His identifications were widely discussed and many of them accepted in the scholarly world.[30] Furthermore he was a gifted landscape painter. In the nineteenth century, a career in cartography often had its origin in the individual's talent for drawing. In van de Velde's case, this was fostered by painting lessons with the famous marine painter Petrus Johannes Schotel at the Royal Netherlands Naval College.[31] So besides map, memoir and travelogue, van de Velde published an album of 100 lithographs, after sketches and watercolours composed during his surveying trip, including depiction of sites in today's Lebanon and Syria.[32] All in all, he expressed his experience of the region in three media – narrative, visual and cartographic – making for three rather divergent images of the same region from the same individual's perspective. Interestingly enough, the striking discrepancy between his three Holy Land images has so far been overlooked by historical researchers and art historians.[33]

Furthermore, the Protestant van de Velde was engaged in humanitarian matters. His first known charitable mission was in fact to the Holy Land. He investigated the state of the Christian population in Syria after the 1860 Druze-Maronite massacres on his second trip

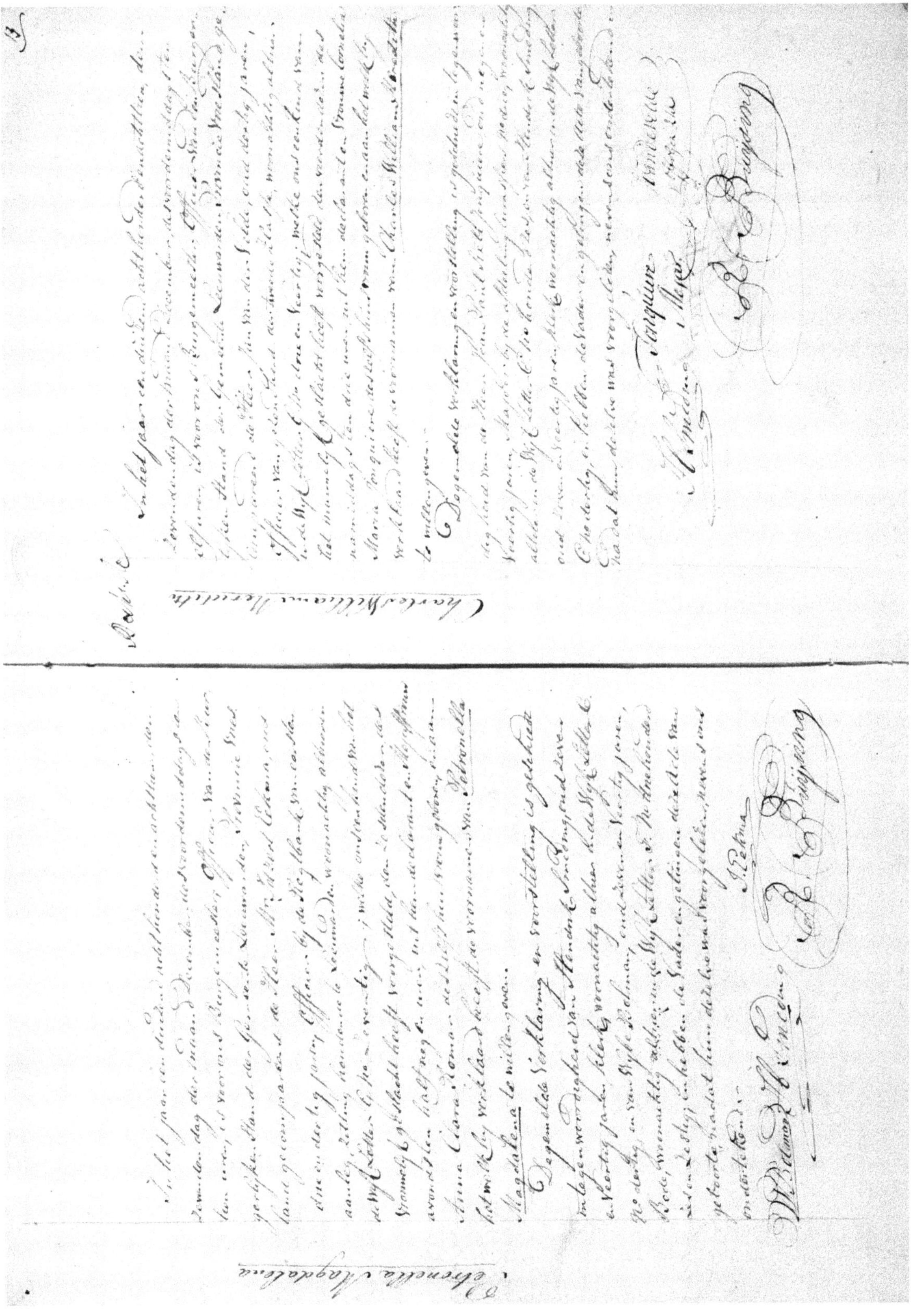

Figure 2.1 Birth entry of Charles William Meredith van de Velde, Gemeentebestuur van Leeuwarden 1811–1914, Geboorteregister 1818, registration no. 3367, record no. 265, suppl. nr. 1–13 (Digitised by Tresoar (Frysk Histoarysk en Letterkundich Sintrum))
Source: Centraal Bureau voor Genealogie

there in 1861–2.[34] One year later, in 1863, he co-founded the Red Cross precursor organisation in Geneva with Henri Dunant and others (see Fig. 2.2).[35] We do not know exactly how he managed to sustain himself financially in all his endeavours. His family was seemingly middle-class with a military background and his father a medical officer with the Dutch Navy.[36] Van de Velde never married.[37] As can be inferred from the decade of correspondence with his publisher, he constantly moved between Geneva, London, Paris, Utrecht and Haarlem, spending his summers in the Swiss mountains. Unlike his correspondence with his publisher, his papers have in all likelihood not survived, nor his map sketches.

Figure 2.2 Charles William Meredith van de Velde, photography, *c.*1863. From the historical album *Portraits des Premiers Adeptes de la Croix-Rouge*, Vol. I (cote AF 681)
Source: Archives of the International Committee of the Red Cross (ICRC), Geneva, Archives CICR (DR), V-P-HIST-E-05504, Geneva, Switzerland

The *Map of the Holy Land* was his second medium-scale cartographic work for the public, after a map of Java (1845, scale 1:700,000).[38] It combined his own eight-month survey of the region with geographic, cartographic and surveying data by third parties. We will elaborate on this way of constructing geographical knowledge based on previous works in detail later. The map was engraved at a scale of 1:315,000 on copperplate and the prints subtly hand-coloured: lakes and coastal lines were painted blue, the borders of the Ottoman *pashaliks* light green and pink, and the route travelled marked in red. It covers an area about 380 km long and 200 km wide, ranging from E 34°30′ to E 37°00′ and N 34°30′ to N 30°55′ (based on the Greenwich meridian), or from the coastal city of Tripoli in Lebanon to the coastal town of Rafah on the eastern edge of the Sinai peninsula, and from the Mediterranean to the Negev desert and the lands east of the river Jordan. A second edition was published in English and German in 1865 and 1866 respectively, whereas the 1858 edition had been in English only. The German version of 1866 also had hypsometric layers in colour that were printed in lithography on top of the copperplate print. This was a recent invention of the Perthes publishing house to create an illusion of the earth's surface relief. Cartographically, the map used a visually attractive terrain shading done by hachures. Dirk de Vries suggested that van de Velde was the first after Guillaume-Henri Dufour, in his *Topographic Map of Switzerland*, to represent relief this way.[39] There are various copies of the map available online (see Fig. 2.3).[40]

One of the many peculiarities of this map is that, except for engraving the plates and publishing it, van de Velde did nearly all the work in person. In the mid-nineteenth century, today's academic subjects had only begun taking shape as professional disciplines and most explorers did not limit their activities to one field. Nevertheless van de Velde's multitude of functions and talents was rather rare, especially the quality of each of them. Equally odd was his 'mapping Palestine single-handed', as an obituary would phrase it almost fifty years later.[41] We have no positive idea how he funded his surveying trip, although we know of his intimate connection to Protestant missions such as the 'London Society for Promoting Christianity Amongst the Jews' (LJS). His surveying trip was rather extensive and led him from the northern edges of the Lebanon ridge, from Beirut, Baalbec and Damascus back and forth across the entire country between the Mediterranean and the Jordan, and led him as far south as the Dead Sea, the Negev desert, Beersheba and Gaza in a zigzag route (see Fig. 2.4). When measuring on location, he got by with a few local aides, guides and interpreters rather than technical assistants. His most reliable assistant, as he put it, was a fifteen-year-old boy who acted as interpreter.[42] Van de Velde was practically a one-man surveying team. Working without trained assistants, he had to use measuring instruments below the best possible of his time. He could not bring a theodolite, which needed several people to operate it.[43] Yet although he measured his angles with a mere 7-inch surveyor's compass, mounted on a tripod, and jotted down the hours travelled between locations,[44] he managed to create a map that was deemed one of the most accurate of the region (see Fig. 2.8).[45]

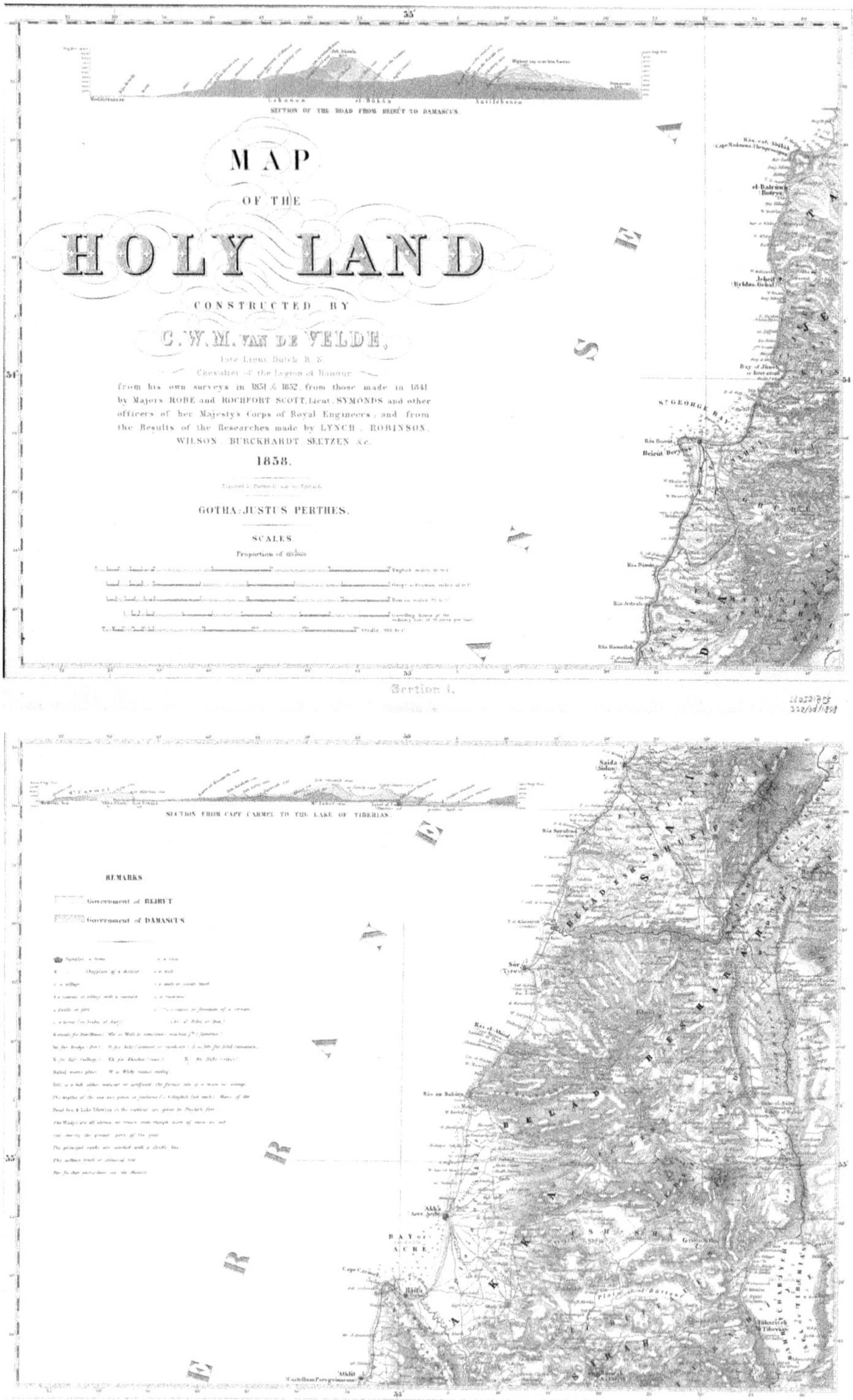

Figure 2.3 C.W.M. van de Velde, *Map of the Holy Land* (Gotha, 1858), sections 1–8
Source: VU University Library Amsterdam, LL.03217gk

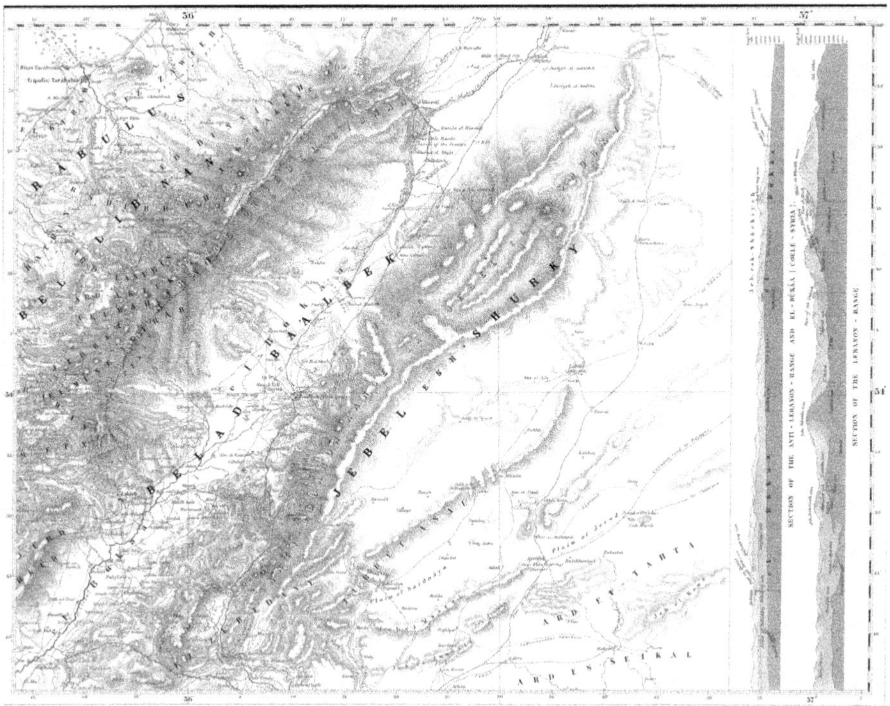
JEBEL ESH SHURKY
BAALBEK
BELAD
ARD ES SEIKAL
SECTION OF THE ANTI-LEBANON RANGE AND EL-BUKAA (COELE-SYRIA)
SECTION OF THE LEBANON RANGE

Section 2.

JEBEL ESH SHEIKH
DAMASCUS
Bahret esh Shurkiyeh
Bahret el-Kibliyeh
Bahret Hijaneh
JEDUR
LEJAH
HAURAN
JAULAN
BATHANIYEH
SECTION OF COUNTRY ON THE EAST SIDE OF THE JORDAN
WEST JORDAN SECTION

Section 5.
Kaisariyeh
Caesarea
Yafa
Japho Joppa
Ramleh
JERUSALEM

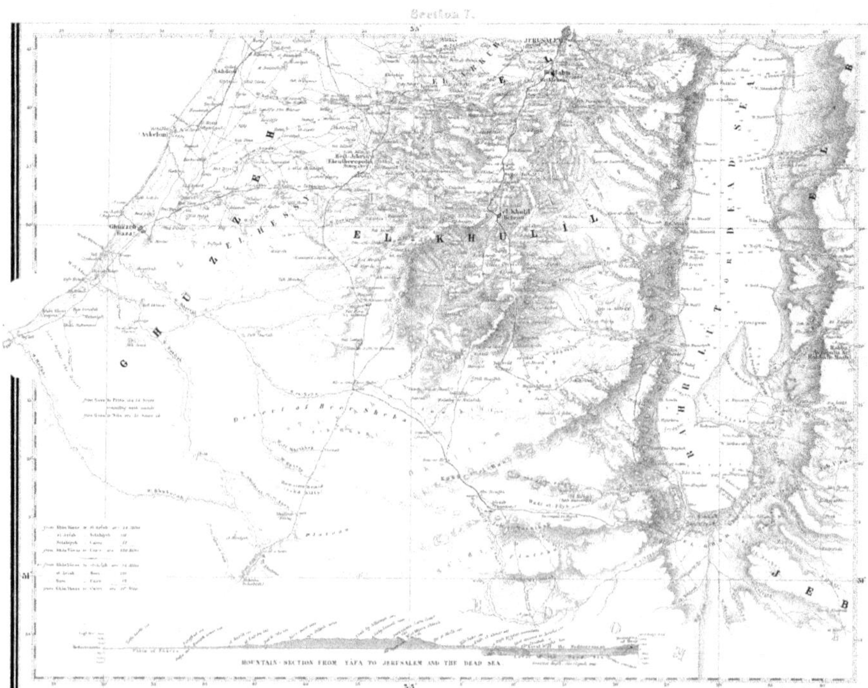
Section 7.
JERUSALEM
Askelon
Ghuzzeh
Gaza
EL KHULIL
DEAD SEA
MOUNTAIN SECTION FROM YAFA TO JERUSALEM AND THE DEAD SEA
Longitude East of Greenwich

Section 6.

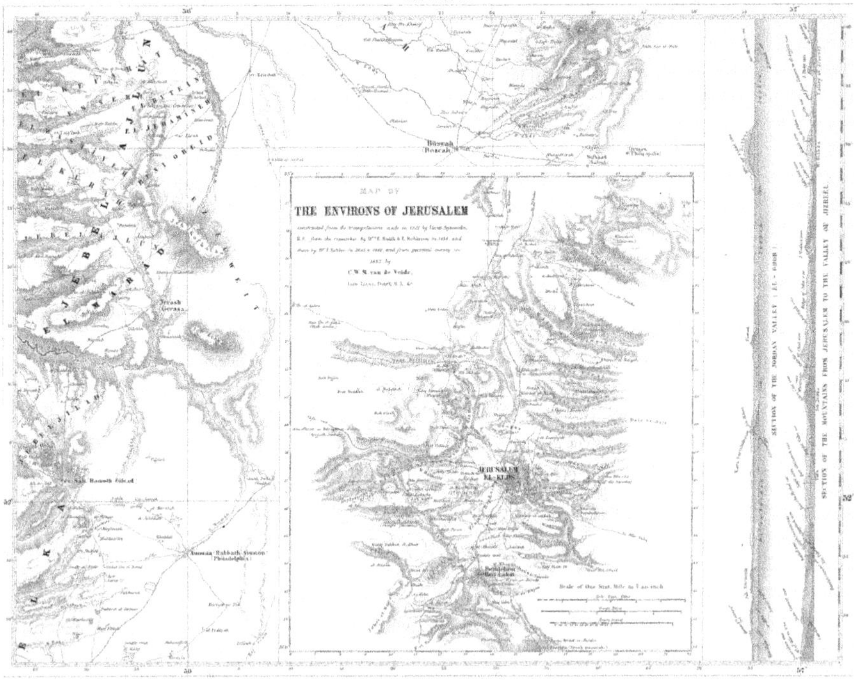

Section 8.

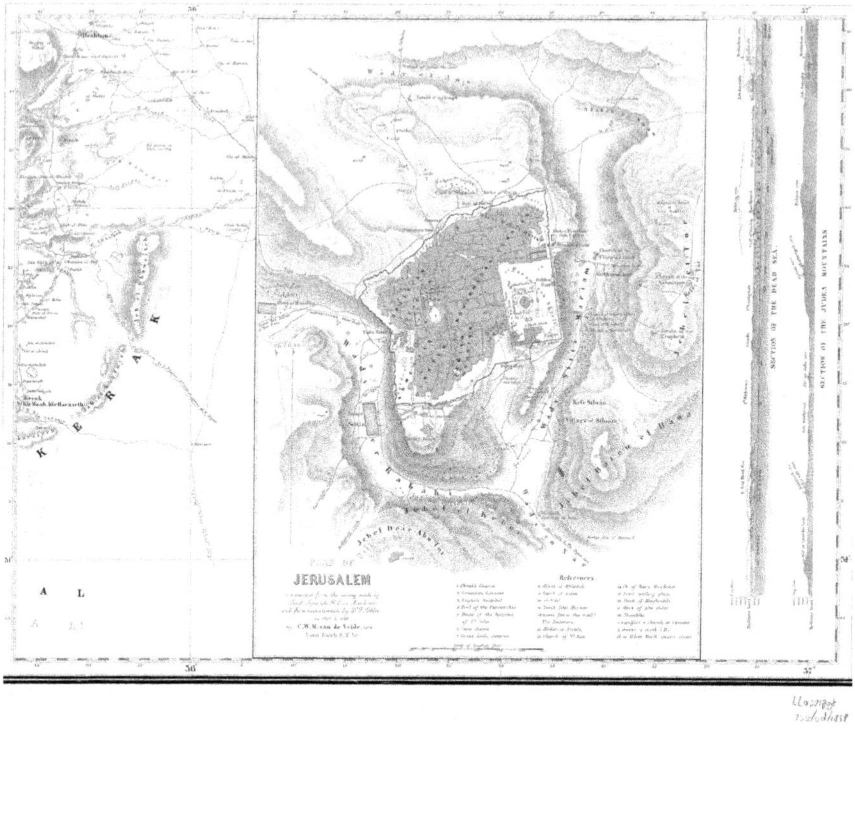

After starting his new job with Justus Perthes in August 1854,[46] August Petermann had put his extensive network to good use for the company and, among others, had written to the Dutchman he had met abroad. We only have van de Velde's answer, which dates from 8 December 1854 and refers to a letter by Petermann of 23 November. Petermann had apparently enquired about the state of works on the map and informed van de Velde that Perthes intended to make a second issue of Robinson's map. Van de Velde congratulated him on his new position, yet had already made arrangements with Blackwood & Sons, the publishers of his English travelogue, as he strongly preferred the map to be published in Edinburgh. Nevertheless, he was interested in a price estimate for having the engraving done in Gotha, as he had not settled on an engraver yet.[47]

Petermann only answered in March 1855. As his note in the top right corner of the letter informs us, he sent the first issue of his new journal *Mittheilungen*[48] and asked about the state of things with the Dutchman's map. Van de Velde's answer came from Edinburgh, where he had, he wrote, met with difficulties which 'make it likely that I shall not publish my Maps & Geographical memoir of Palestine in this part of the world'.[49] He wondered if Perthes would be interested and if they could secure the copyright in Britain, which was a matter of utmost importance. Petermann must have suggested meeting in London, as the next letter of June 1855 originates from there. Upon Petermann's recommendation, van de Velde wrote to Bernhardt Perthes, a third-generation publisher and head of the Perthes publishing house in Gotha, capital of the Duchy of Saxe-Coburg-Gotha. Although the Dutchman had prepared his map in English as he planned to publish it in Britain, he assured Perthes that this would suit the audiences on the continent as well. He proudly pointed out the value of his map, which was based on his own surveys and the unpublished triangulations of a group of British Royal Engineers from 1840 and 1841 respectively.[50]

These British triangulations were the second modern attempt at surveying the region.[51] The first ever was initiated by Napoleon, whose campaign in Egypt, Palestine and Syria (1798–1801) not only laid the foundations for the Oriental craze in nineteenth-century Europe but also paved the way for a scholarly Western appropriation of the Levant. Mapping, which was becoming a key instrument of colonial power, was an integral part of the Egyptian Expedition. But the time Napoleon spent in the Mediterranean was much too short to accomplish a full triangulation. The map issued as part of the *Déscription de l'Égypte*[52] was praised for its beauty but lacked accuracy. Its greatest flaw was that details were added to the map *ad libitum* where the French had not been able to survey, which resulted in erroneous geographical features. Despite its shortcomings it was a first ground-breaking attempt at a modern, triangulation-based map.[53]

The Oriental Crisis of 1840, an armed conflict between Egypt and the Ottoman Empire, brought European powers yet again to the shores of Palestine. The Ottoman governor-general of Egypt, Muhammad Ali Pasha, had gained so much power that he threatened to overthrow the local regime of the Ottomans. England, Prussia, Russia and Austria aligned to support the young Turkish sultan Abdülmecid I, whereas France supported Mohammed

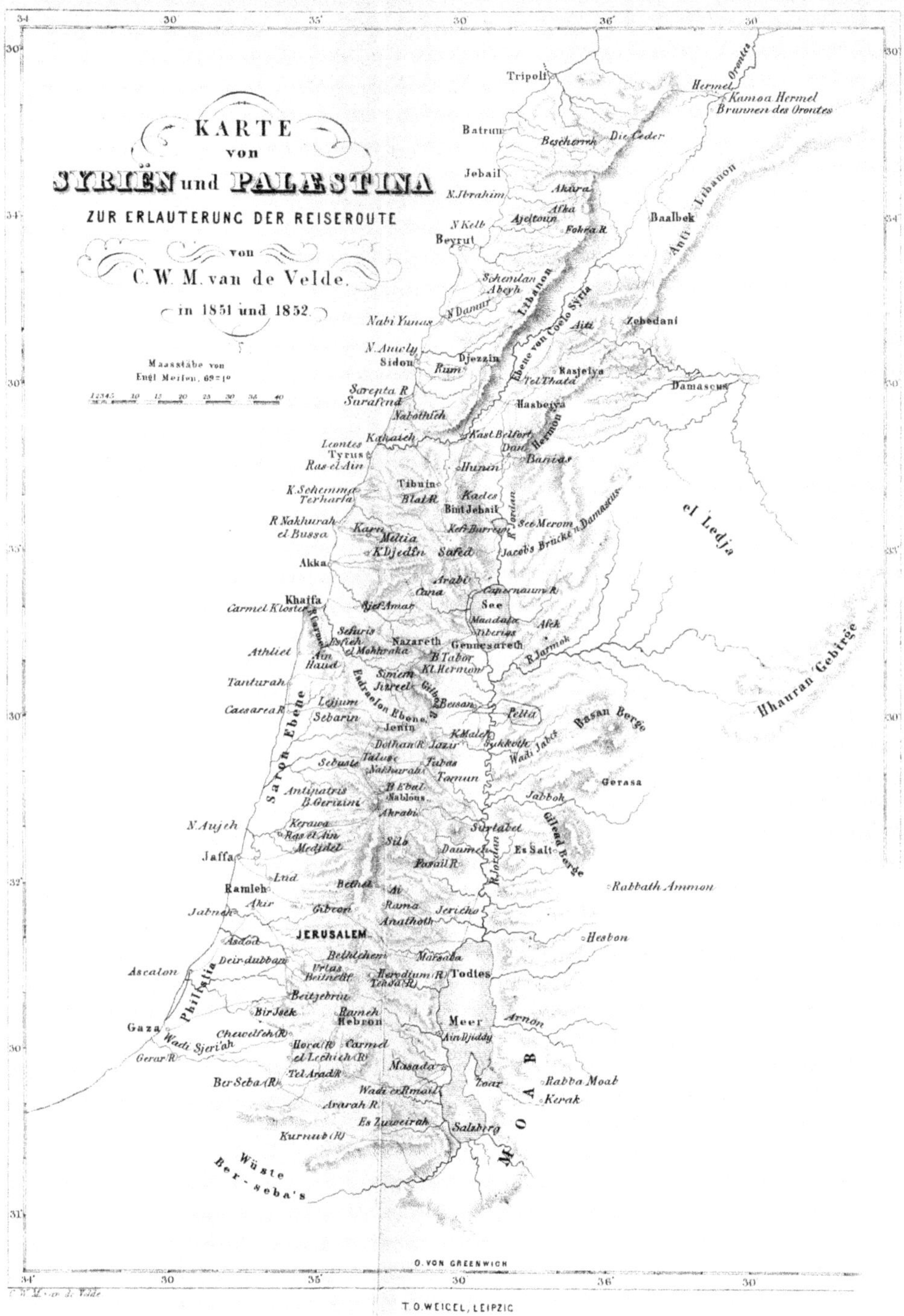

Figure 2.4 'Karte von Syriën und Palaestina. Zur Erlaeuterung der Reiseroute von C.W.M. van de Velde in 1851 und 1852' (Map of Syria and Palestine. Illustrating the Travel Route of C.W.M. van de Velde in 1851 and 1852)
Source: van de Velde, *Reise I* (Leipzig, 1855), FBG SPB 8° 3030.41 (1)

Ali. When British ships besieged the coastal town of Acre in November 1840, a group of Royal Engineers sappers and miners seized the opportunity to survey the coast and the hinterlands. In various excursions, they made it as far inland as Jerusalem. Yet the resulting maps and triangulations were never published in their entirety. C.W.M. van de Velde – probably because of his own navy background – managed to obtain this valuable information from the Board of Ordnance in London, albeit only after he returned from his own surveying trip. As much as van de Velde asserted that his survey had only been 'verified' by the British officers' survey, he was able to fix his own bearings within their triangulation network, measured with the technically superior theodolites of the British Royal Engineers.[54]

August Petermann wrote to his employer:

> I have just met with Lieut. Van de Velde who showed me the first draft sketch of his map of Palestine. The same is bordered in the South by Gaza and in the North by Tripoli, is 53 English inches high and circa 38 inches wide (the outer margins). Lieut. Van de Velde wants to divide the map into 8 sections at a scale of 1/450,000. Concerning the value and correctness of this map, I can only express my opinion by saying that I would like this map to be executed and published by our publishing house.
>
> My estimate of production costs is about 2000 to 3000 Thaler. It seems that the 8 sheets would match the Handatlas[55] size approximately.
>
> Concerning the conditions under which van de Velde would give the work to you, he either wants royalties that would equal the value of his finished sketch, or keep his share instead and receive royalties for the number of copies sold.
>
> If you think it probable that you would produce and publish the work, van de Velde would come to Gotha himself immediately to show you the drawings and have a chance to discuss matters in detail with you. Hence please have the kindness to send us your answer as soon as possible, as Lieut. van de Velde's moves altogether depend on your answer.
>
> NB. Lieut. Van de Velde's map will be a renewal of Prof. Robinson's map which is to be published by Dietrich Reimer – like 'Mittheilungen' and 'Zeitschrift'.[56]

Petermann enclosed a letter by the Dutchman, full of confidence: 'It is unnecessary to prove the importance of my map to you.'[57] Four days later, 'with great delight', the publisher accepted the offer to publish the map. In July, van de Velde came to Gotha to sign the contract and draw his map for the engravers. The length of his stay is unclear, but it was long enough for him to become friendly with most of the people involved in the map-making process at Justus Perthes Geographische Anstalt. The contract signed on 23 July 1855 is part of the correspondence. Herein the cartographer was assigned responsibility for delivering a ready-drawn map for the engravers and supplying the print-ready text of the memoir to go with it. The map was to be published in English and German and exclusively distributed by Perthes. The publisher would cover the costs of engraving, printing and colouring the map, as well as for paper, book binding and all expenses related to sales and distribution. Both parties agreed not to make any other map from this one, be it national or international, yet the publisher was granted the right to use the cartographic data for other products. Two-thirds of the net revenue after production costs would be credited to the cartographer yearly on 1 January. Both men signed and sealed it with their family crests.

Van de Velde got in touch from Switzerland four weeks later, in August 1855. There he had met with Titus Tobler, a Swiss Palestine explorer and physician by training,[58] to make a new map of Jerusalem based on Tobler's own measurements. The Jerusalem plan of the Royal Engineers Aldrich and Symonds, also from 1840–1, had erred in one crucial detail – the wall of the Haram El-Sharif or Western Wall. Through this error, the entire city had been positioned too far west on their plan, which van de Velde had intended to use for the small Jerusalem plan to be inserted in his map.[59] With Tobler's help he had constructed a new plan of the city and offered Perthes to develop it into a full map to be published separately. He closed with best regards to Petermann and to Mrs Perthes, whose kindness he remembered fondly, as he said. From now on almost every letter would contain warm regards to members of the publishing house and their families, a sign of the personal ties created when van de Velde had visited Gotha.

Bernhardt Perthes immediately agreed to publish the Jerusalem map as well; the ensuing correspondence with Titus Tobler fills another bulky folder. Tobler's and van de Velde's map of Jerusalem was published in 1858, too,[60] but well ahead of the *Map of the Holy Land*, as we will see. Every now and then, the relation between the two very different scholars surfaced in the correspondence. Despite their good cooperation and mutual respect, van de Velde could not withhold sarcastic comments about the nit-picking Swiss 'savant', who he once described as a bit lengthy and finicky, like his books ('tout comme ses livres un peu long et penible.')[61]

THE MAKING OF A MAP

In the meantime in Gotha, the map production had begun. It was September 1855 and van de Velde was waiting for news from two men, Captain Layard and J. L. Porter. Henry L. Layard, secretary of the LJS (not to be confused with the famous Assyriologist), had been an ardent supporter of van de Velde's cause, probably also financially.[62] Now he had volunteered to print an English prospectus to promote the map in all possible ways. The Irish Presbyterian minister Josias Leslie Porter had been sent to Damascus as a missionary to the Jews in 1849 and travelled the region extensively.[63] Porter's *Five Years in Damascus*, officially advertised for and eventually imprinted as published in 1855, was not yet available, much to the dismay of the cartographer who needed the accompanying map to complete his own. This missing link, which covered a region van de Velde had not been able to survey due to lack of time, delayed the printing of the map for months. Other sources of information were of great interest as well. He enquired of Perthes if Heinrich Kiepert, the Berlin cartographer of Edward Robinson's and Eli Smith's Palestine maps, had already visited Gotha and what news they had been able to gather from him.[64]

Over the winter, van de Velde relocated to Geneva to make his watercolours into lithographs. There he received the first proofs of the Gotha engravers' work. The lengthy and troubled correction process that ensued would take a lot of time and effort. Eventually it would put the relations between cartographer and publisher to a test. First of all, van de Velde demanded his sketch map to be sent along with the proofs in order to compare both. Other than that he had already found various mistakes: misspelled names; names of rivers and *wadis* printed in different sizes and typefaces, mountains' names that were too large, etc. He had agreed upon the duties for the Jerusalem map with Tobler: 100 Thaler for Tobler, who also wrote a short text to be published with the map, and 200 Thaler for his own cartographic work.[65]

Van de Velde worked at the Geneva studio of the famous Alexandre Calame.[66] He had arranged to meet Porter in Geneva and asked Perthes to wait for the information the former would supply. He had also written to the Dutch ambassador at the Sublime Porte and to missionary William Thomson[67] at Sidon for details of the borders of the Ottoman administrative districts. But Porter became ill and could only offer to have his publisher John Murray send a copy of the map in the following month, when it was expected to go into print. This meant another unpleasant delay.[68]

In London, Layard had the prospectus ready for distribution. Van de Velde sent Perthes a list of people he had addressed for promoting his map. These can be roughly divided into three groups:

- members of the Evangelical Alliance, an organisation 'to promote religious freedom', founded in England in 1846 (Rev. Edward Steane, Sir Culling Eardley, John Henderson of Park, to whom van de Velde had dedicated the English version of his travelogue);
- ministers and missionaries (Louis Frederick Liesching, Ceylon; Andrew Murray, South Africa); and
- diplomats (Count Zuylen van Nijevelt, Constantinople).

But this is not to mention the crowned heads of the Netherlands, Belgium, Wurttemberg and Prussia; the latter two were to be contacted by Perthes directly. At the end of 1855, Porter's book had still not come out and van de Velde was stuck in Geneva working on the lithographs. Because the local libraries did not hold the literature to finish his map memoir, he asked Perthes for books.[69] This was not the last time his three media, the map, the books (travelogue and memoir) and the album, proved difficult to coordinate. He was too busy to go to Gotha in spring, he wrote, because he had to promote his Palestine album in Britain. Once the nomenclature on the map had been corrected, the engravers could easily continue with the terrain, he said. Concerning the Jerusalem map, he had just received very time-consuming corrections from Tobler.[70]

Time was running short. The prospectus had been sent out as far as South Africa and Ceylon, but not even the engraving had been completed. Perthes pressed for results. Van de Velde's answer of January 1856 is typical of the Dutchman's temperament. He assured Perthes that he was doing everything he possibly could, but without Porter's map he could not proceed. His health was frail and he was working hard; more pressure might even prolong the map's completion. Perthes simply had to be patient. Then van de Velde declined an offer that, unlike other letters, did not survive in copy. It seems that Perthes had suggested a permanent position as resident cartographer to him. Much to his dismay, the Dutchman wrote that he could not accept this fine and flattering proposition.[71] It would be disappointing if van de Velde moved to Gotha only to have his health fail after six months, he added, and claimed that even after the recent period of rest he felt that too much stress might render him incapable for a long time. As a result, he could only take on a light amount of work ('tout doucement'). Overall, his delicate state of health forced him to resort to art rather than geography and he considered it his duty to follow the path thus pointed out to him by the Lord, he wrote. If God granted him the health and strength and if he refrained from hard work for a while, he might be able to devote at least part of his time to cartography.[72] He sent Tobler's corrected Jerusalem map back to Gotha three weeks later, in February 1856, still complaining about the unpleasant geographical work. At least he could start completing the section that relied on Porter's map, now that the book had arrived. From London, Captain Layard had informed him confidentially that Palestine had become an important subject among high-ranking people in Britain. After the anticipated peace with Russia (the Crimean War was ended by the Treaty of Paris on 30 March 1856), the fate of the Holy Land would be one of the most pressing issues in Europe. Layard would promote the map in the LJS journal. Also the Dutch ambassador at the Sublime Porte, Count Zuylen van Nijefelt, had voiced great interest in the map and offered his support. Finally, things started looking good for the *Map of the Holy Land*.[73]

Then everything fell apart just two weeks later. In mid-February 1856, proofs of the engravings arrived in Geneva, this time including lines, names and terrain all positioned within the scaled map frame. Van de Velde was in shock: the engravers had ruined his map by reducing it in size and corrupting his graticule! This was a disaster for him. The proportions and terrain were contorted; latitudes and longitudes were not consistent with the already finished geographical tables for the memoir. He was speechless, or so he wrote to Perthes. If this was a map based on hypothesis, he might have accepted it. But he had measured all the angles himself! This map would be the base of many future maps! Had they thought a map was elastic fabric and could be distorted at will? He had placed complete faith in Perthes, but this was a serious situation.[74]

Bernhardt Perthes's answer has only survived in an almost illegible draft. Yet it is important for understanding the technical problems inherent in the map-making process as it was customary in the nineteenth century.[75] The transfer of the original design, drawn on tracing paper by the cartographer, to copperplate by the hand of the engravers was a crucial moment. It could also go wrong, as seems to have happened here. As can be reconstructed from Perthes's answer, van de Velde had supervised the engraving of one section himself when he was in Gotha, then handed over the other seven as sketch maps for the engravers Hermann Eberhardt and Heinrich Stichart to complete.[76] This task had obviously not been carried out to his satisfaction. The sharp accusations hurt the publisher's pride, as his detailed reply shows. The smaller size of the print as compared to the original sketch, Perthes said, was due to paper shrinkage. Maps were printed on wetted paper and, since paper was made from natural materials, namely cellulose and rag fibres, it could shrink in the drying process. Yet it seems unlikely that this could have accounted for such a severe distortion of the grid. After all, printing maps was the core task of a map publisher. Perthes added the engraver's reply as well. His vindication hints at the technical problems inherent in transforming a map from sketch to copperplate:

> The arrangement of my grid is entirely correct and has been devised exactly according to the north-eastern [copper] plate. Yet it is a regrettable mishap that the dimension of the grid does not fully coincide with the [map's] projection. Due to the fact that the north-eastern sheet was drafted first and alone, the construction of the grid on the other plates had to conform to the dimensions that had been determined there. The deviation only became apparent when the blueprints were applied to the finished grid on all plates and the matter could not be rectified anymore. Incidentally, the dimensions are not entirely correct on the original sketches neither, most likely due to the fact that lines for parallels and meridians were added only after the map had been drawn.[77]

The sketch they had kept in Gotha, Perthes asserted, agreed with the engraving except for minor details. The engraver had assured him that the deviation of the other sections could not be any more serious. Moreover, the engraver said he could not believe that the southern plate deviated by more than a quarter minute in Gaza. Yet mailing back and forth was of no use, Perthes declared, so he asked van de Velde to make all necessary corrections and specify how exactly the graticule needed to be changed in order for the map to satisfy all scientific needs. He could rest assured that all revisions requested as absolutely essential, Perthes wrote van de Velde, would be executed diligently.[78]

Van de Velde was disgruntled. Presently he could not discuss the issue, he wrote back on 8 March 1856, because it had left him too agitated. He felt vindicated that the engraver had been careless and now pretended nothing had happened while the map had indeed been 'mutilated'! Van de Velde was crying, so he wrote, over the fact that this had happened to him after so many years of trouble and work on the map. His sole consolation was that this had come about without Bernhardt Perthes's knowledge, in whom he placed his full trust. He would send Perthes all relevant materials in order for him to judge matters himself.[79]

There were more problems ahead. At the end of March, van de Velde sent back the map with place names corrected. When incorporating Porter's map data, he told Perthes, he had found that Porter had made a mistake. Instead of correcting the compass declination by 8¼ degrees north-west, he had corrected it by 8¼ degrees north-east. Consequently Porter's map erred by an angle of 16½ degrees.[80] The magnetic declination, or variation, indicates the difference between magnetic north and geographic north. Geographic north, the meridian that ultimately points to the North Pole, is a construction of geographers, whereas the magnetic north to which the compass needle points follows the direction of the earth's magnetic field. Both are not fully congruent. Moreover, the earth's magnetic field changes over time, so compass bearings have to be adjusted in order to make maps that correspond to the conventions of geographic north. The magnetic declination is measured by referencing the celestial poles such as the North Star or Polaris in the northern hemisphere and has to be updated frequently to keep up with the earth's changing magnetic field. The error forced van de Velde to construct Porter's map anew, which strained his health and patience once more. Yet at long last he was content to have found the mistake and subsequently completed the last section of his own map. Concerning the administrative borders, even the Dutch ambassador at the Sublime Porte had not been able to identify them with certainty. Apparently they were subject to continuous change, to the extent that the governors themselves did not know where their territories ended, he wrote. Hence he had not drawn border lines but only given names of districts and divided the country in the two *pashaliks* of Beirut and Damascus. Fittingly, antique place names were to be printed in Antiqua font. All other necessary corrections had been made directly on the proof map that he sent back to Gotha.[81]

Despite the criticism, van de Velde ungrudgingly acknowledged that the engraving of letters and the terrain shading were very beautiful. But then he could not but come back to the issue of the distorted grid: If not for the deviation of latitudes and longitudes from his manuscript, he would have gladly accepted the map, he wrote. He had thought of coming to Gotha himself to settle the matter, but his travels sent him elsewhere. And actually the mistake was more than obvious: Perthes only had to take a compass and compare the manuscript map and the engraving himself.[82]

It is not easy to retrace this dispute more than 150 years later. A point of great deviation mentioned by van de Velde is the location of Gaza. The sketch letter by Perthes gives its latitude as 30°30′N. In his later detailed letter, van de Velde claims the deviation in latitude to be more than half a minute; he huffs at the engraver who 'did not believe that'.[83] The memoir lists Gaza's position as 31°29′45″N (and 34°33′10″E, but only the latitude was questioned). In section 7 of the finished map of 1858, Gaza sits a little south of 31°30′N, so its location matches these coordinates. Generally latitudes and longitudes are indicated on the map frame alone; only full and half degrees are drawn as lines across the map. It thus takes at least a long ruler to read the coordinates of individual places from the map. But, even when examining the map closely, the location of places on the map corresponds with the coordinates in the memoir.[84]

Beyond technical problems, this was a question of trust between van de Velde and Perthes. It was also a question of the cartographer's honour. He had measured the angles himself and vouched for their correctness. On the Gotha side, the publisher and the engraver were offended by the allegations. Van de Velde appealed to Bernhardt Perthes's 'Christian integrity' to save the honour of his work, with agitated handwriting and words spread widely over the page:

> I reassure you that my confidence in you has not failed, and it is with this confidence that I call upon your heart and conscience. [...] I will not embarrass you, we must respect each other mutually as citizens and, what is more, as disciples of Jesus Christ. I do not intend to stray from my way by causing you trouble. You must respect me as much as I respect you. That is all I have to say.[85]

But there was good news. Van de Velde's Palestine lectures had met with great applause in Geneva. Not only had the newspapers flattered him, but he had been awarded the bronze medal (very rare, he added) of the 1835 Calvin jubilee. Through his lectures he had already received several subscriptions. From England he heard that Williams & Norgate had received subscriptions beyond expectation for a map not yet in print. He would see for himself, as he was leaving for London soon. Apparently van de Velde spent considerable time acquiring subscriptions for the map (and his album, too). His letters mention the subject over and over again, forwarding names and addresses to Gotha. As we have seen, the contract remunerated him per map sold, so naturally he wanted to promote the map as much as possible. He even asked Perthes to send him two copies each of the proofs, so that he could show them to people interested in the map.[86] Lastly, after a long list of rants, van de Velde could not help gloating over Porter's mistake. Porter had still no idea of his error; he wondered if he should inform him or not. He felt it would be sad for Robinson's map if Kiepert did not notice that mistake and asked if Perthes had heard anything about the latter's map yet.[87] This kind of ironic side blows at competitors, fellow Biblical Geographers and others can be found throughout the correspondence.

In summer 1856, van de Velde was in London for six weeks, staying with Captain Layard of the LJS. He then travelled around England (Plymouth, Bristol, Exeter, Cheltenham) to promote his work at meetings of geographical, biblical or missionary societies. He went on a promotion tour like authors still do today. Williams & Norgate, the prospective English distributor of the map, had already gotten 80 subscriptions; he had received a dozen more since his arrival in London. In contrast, he had the greatest difficulties finding subscribers for his album, although he had had the honour of showing it to the Queen and Prince Albert.[88] After returning to London in July from his country tour, he presented his work at the RGS. He had hoped in vain to receive the proofs from Gotha in London. Also he regretted not having met Petermann, who was apparently in London at the same time, but the city was big and they had both been busy, he said. Meanwhile he was planning to go to Gotha in late August, if Perthes agreed.

After leaving London, van de Velde stayed with his sister in the Dutch city of Haarlem. Madame J. van Coppenaal was widowed (referred to as 'veuve') and apparently very close to her brother. He often travelled with her and her children and had planned to take them to Gotha for a summer retreat. Now Perthes had asked him to come only in mid-September, causing them to miss this vacation. He was annoyed that the engravers were behind schedule but wondered what Perthes thought about the watercolours. Would there be a chance of selling them in Germany? His hopes were not too high as a copy of the album cost 250 francs.[89]

The journey to Gotha had to be postponed once more. Shortly before he was scheduled to go, another letter advised van de Velde to come in December only. Moreover, according to Bernhardt Perthes, the German market was not promising for such a costly album, whereas in the Netherlands van de Velde had received a decent number of subscriptions, he wrote, and the Dutch king had accepted the dedication. Next, van de Velde planned to go to Paris to find a publisher for the album.[90] He had to stay there for two months to supervise the lithography, he wrote in November 1856, so going to Gotha in December would be a great inconvenience. He did not want to have the work on the map wait but his album was equally important. In Paris, he had also reclaimed the copperplates of his Java map. They had spoken about the matter and he wondered if Perthes would want to buy the rights to this map. If so, he would bring the plates to Gotha.[91]

Perthes sent the map proofs to Paris and van de Velde promised to do the corrections as quickly as possible. Yet Robinson's book had just come out, so he not only had to compare the proofs to his sketch map but also to Robinson's work. Indeed, *Later Biblical Researches*, the result of Edward Robinson and Eli Smith's second journey in Palestine in 1852, was published in 1856.[92] Van de Velde had actually met the two men and travelled a part of the way with them on his trip.[93] In the same letter he also wrote that he had gotten his hands on another valuable map: authorised by Lord Clarendon (Secretary of State for Foreign Affairs), he had received a copy of a map for use of the Foreign Office, made by Rochfort Scott, member of the Royal Engineers team of 1840–1.[94]

The first instalment of his album was due out in early 1857. After that he would be free for the map and the memoir, he wrote, which also needed corrections after Robinson's *Later Researches*. While he was waiting for the album, he asked if Perthes could send him Stanley's and Porter's works.[95] Self-assuredly, van de Velde added that, on seeing Robinson's book, Perthes would notice that the maps were executed 'in a manner unable to harm our [map]'.[96] This probably referred to the size – van de Velde's map was to be published in a scale of 1:315,000, the map to accompany Robinson's book was a mere 1:800,000.[97]

In March 1857, he was busy updating the map with the new information found in Robinson's latest book. He even had his original material sent from the Netherlands. A slip of paper with a detail to be corrected has been preserved with this letter and hints to a general question of cartography:

> Often an important city in antiquity is nothing but a small village or ruin today. In this case, what is insignificant should be printed small and what is of importance in bigger letters; for example Hesbân vs. (**Heshbon**).

Here he drew the two places' names in fine hand lettering, the current one smaller, the antique one bold and in brackets.[98] In the end, he was ready to send his corrections to Gotha:

> Here is the map finally, completed to the highest possible degree of perfection. It is correct that these additions add to the cost of engraving, but it is also true that the map will have gained much value through them. Yet after the publication of Robinson's book, my map had to be updated with his work. Now the map does not have to fear any competition by Robinson's.[99]

All that remained to be corrected was the sheet depicting Gaza to Jerusalem and of course a final revision of all sheets. Van de Velde had no idea how Perthes was going to fix the serious error of overall diminution of the map and the division of its border, but confided in him completely. He was ready to go to Gotha for the final revisions. Then they could also discuss the plans to make a travellers' map of Palestine and a Biblical atlas.[100]

And really, in late April 1857, almost two years after their first contact, van de Velde wrote from Paris that he could leave for Gotha in May. Another notice from Utrecht informed Perthes that the Dutchman would arrive in June only; he had been ill with fever. He was also going to bring the next instalments of his album *Pays d'Israël* (see for example his view of Jaffa, Fig. 2.5).[101] At this point in the making of the map, we face the drawback of relying solely on archival material. Since we do not have any first-hand account, we do not know what happened after van de Velde went to Gotha in person and how the problems that had been discussed in dozens of previous letters were eventually solved. But they were solved, and never came up again.

Van de Velde's next letter dates from mid-October 1857, again from Paris. As it mentions a stay in the Swiss mountains (where he worked on the memoir instead of enjoying nature, as he said), he most likely did not spend the summer in Gotha. Work on the memoir was more time-consuming than anticipated and would take a few more weeks at least. If God granted him strength and good health, he hoped to put the corrected manuscript in Perthes's hands around the end of the year. The map was almost ready to go into print.[102]

After that, the letters by van de Velde bear margin notes about the content of replies sent, usually on the top right of the first page. In this case, the note said '4/11 Shall send Memoir by end of year or map will be published without'.[103] These were by Adolf Müller,

Figure 2.5 'Yâfa (Japho, Joppe). Vue Septentrionale.' (Yafo. View from North) Lithography from the album van de Velde, *Pays d'Israël* (Paris, 1857), no. 56
Source: Private collection

one of the publishing house's executive directors, as can be inferred from the handwriting. Müller and his colleague Rudolf Besser took over the management of Justus Perthes Geographische Anstalt because on 27 October 1857 Bernhardt Perthes had suddenly and unexpectedly died of typhus at the age of 36.[104] From now on, van de Velde corresponded with Müller, who took over all responsibilities in map-making, just like his employer had done before. And, like all other authorised members of the firm, Adolf Müller signed his letters simply as 'Justus Perthes', as a representative of the company. Only occasionally do the letters allow a glimpse of the person behind the business, such as when van de Velde got on the manager's nerves too much.[105]

The death of his business partner, a like-minded Protestant,[106] came as a great shock to van de Velde. They were so close to printing the finished map that he felt they should continue in the spirit of the deceased. The memoir would be ready by the end of the year, but van de Velde's head was weary, he wrote, which slowed him down. It would be a great disadvantage if map and memoir appeared separately. But since he continued adding the latest information the map would not be outdated by the time it came out. He might send a few more small corrections, but after that they could start printing. He wondered if Perthes publishers had decided on printing the map and memoir in both English and German.[107] As can be inferred from a margin note, Gotha had decided on an English map and only the memoir in both languages. Indeed, from these margins we can learn a lot about the publisher's part of the map-making process. In late November 1857, Perthes (represented by Adolf Müller) sent van de Velde one set of the fully corrected map, all sheets included. In return, van de Velde was supposed to send as much of the memoir as he could in order to have it typeset in time. Perthes also informed him that they had been unable to find subscribers for his album.[108]

All parties were eager to grant the printing approval. The pace of events picked up with the proofs being sent back and forth by imperial messenger 'à grande vitesse'. All the same, van de Velde asked Müller to write to him in private about the Perthes's family and his own. They had, after all, become very dear to him, van de Velde wrote; Bernhardt Perthes's death would have a lasting impact on everyone.[109] In late January 1858, he sent the final corrections of the map and gave his print approval. He added instructions for the colourists, who were supposed to add only a very light touch of colour. Bold colour would give, as van de Velde put it, a rather common impression ('Une couleur trop forcé donne un aspect commun').[110] The memoir was almost ready as well. Only the chapter on biblical names was missing, but he hoped to send it a week later. Because he was suffering stress, van de Velde started complaining again. The memoir gave him sleepless nights and made him neglect his other work, he wrote. He had ruined his health for this map and did not understand why Müller had to constantly pressure him. It was only through God's grace that he had been able to work in the last days and now the memoir was finally finished.[111]

This map project would not have been the same if the final stretch had gone smoothly. Van de Velde could not clear the memoir for printing because the English version was full of mistakes. Breitkopf & Härtel, an old and prestigious Leipzig printer and publisher, was in charge of typesetting. Apparently English was not easy for them to deal with. Van de Velde had found an Englishman in Paris who would correct proofs, and according to him they contained countless mistakes. But there were technical problems, too. The memoir was 360 pages. Usually 16 pages were set and mounted on one print sheet, 8 on the front, 8 on the back. This step in prepress printing is called *imposition*, a standard procedure to minimise the use of paper and press time. Yet it proved a stumbling block for setting a text whose author was in Paris working on his lithographies. Sheet by sheet was typeset in Leipzig, proofs printed and sent to van de Velde, who sent them back to Gotha with his corrections. From there, they were sent to Leipzig for the corrections to be implemented. This took weeks for the first sheets. While Breitkopf & Härtel ultimately managed to set three sheets in a week, due to the nature of letterpress imposition these did not contain consecutive pages. If there were many mistakes, it was entirely possible that the first batch of already corrected sheets had to be adjusted to the new corrections and sent to Paris again. It soon became obvious that the English memoir would not be ready for print anytime soon.[112]

The memoir turned out to be the roadblock in the entire map-making process. Memoirs were no simple by-products but a central part of nineteenth-century cartography's scientific standards. They provided detailed information about the map's construction, thus both enabling others to retrace its making (as van de Velde had done with Porter's map) and documenting the cartographer's sound scientific work by providing the data that had been implemented in the map in the form of numbers. For the geographical positions in section 5 of his map alone (from the Yezreel Valley to Jerusalem), van de Velde listed no less than eighteen major sources, not including his own measurings.

In short, map memoirs asserted the map's scientific value by making its production transparent. They also provided others with the raw data used in the map's making, like today's publication of research data together with research results. This went hand-in-hand with fundamental changes in the sciences, namely an increasing appreciation of empirical work and a critical assessment of sources in the historical field.[113] Yet, while Perthes excelled at being an international publisher of the latest geographic information, doing so in multiple languages still presented a problem. Competence in English was not available in-house, nor were the Leipzig typesetters fully able to handle it adequately.

Almost three years had gone by since van de Velde first got in touch with Perthes. The map was ready to go into print and any delay put it in jeopardy of being outdated. A heated debate ensued about priorities between Müller and van de Velde; the map over the memoir and timeliness over quality. Müller urged van de Velde to stop insisting on a final revision; they would do this in Gotha. Now Müller was experiencing the same thing as his late employer Bernhardt Perthes before him: In a lengthy emotional rant, van de Velde made clear how important and valuable the memoir was for the entire map, but also how much he insisted upon quality and perfection until the very end. For him, this was the fruit of seven years of hard work, for which he had sacrificed his health and money. For Müller it was just another map but, for him, it was his baby. And, as he sometimes did when he wanted the other person to understand, he added a few lines in German: yes, it was true that Müller wrote him all this in friendly words. But, likewise, he had to believe van de Velde's words, even if it was difficult. Forcing him to abandon his brainchild was not worthy of the noble Perthes's establishment. Van de Velde had managed to find two Englishmen who were now jointly correcting the proofs in Paris. Perhaps he could start sending back the second half soon. His health was frail again, he added. He much deserved a summer vacation in the Swiss Alps with his sister and her family. Müller and his wife were cordially invited to join them to recover and gladly look back on the hardships mastered together.[114]

The frequent mentions of weak health, mixed with other lamentations, raise the question of whether he really suffered lasting injuries on his surveying trip. In his travelogue, he recounts a fever towards the end of the journey.[115] Quite tellingly, only leaving the land on a boat could stop this illness, at least in his narrative. In the letters to Gotha, he occasionally mentions a few days of fever.[116] We also find mention of his fragile health in the letters by fellow Palestine traveller Titus Tobler.[117] It seems very likely that van de Velde had contracted a benign form of malaria (*tertian* or *quartan* malaria) on his surveying trip, a disease that was common in the region, and now suffered from the typical recurring fevers. Surveying the Holy Land, if it did not cost one his life, was still a physically strenuous, sometimes dangerous project.

In May 1858, one of the English copy-editors in Paris fell ill, the other had more urgent work to take care of. At the end of the month, van de Velde wanted to leave for the mountains. He again invited Müller and his wife to join him. To speed up corrections of the difficult place names, the typesetters could use Robinson's *Later Researches* and van de Velde's map as reference.[118] In June, the Perthes publishing house finally started corresponding with Breitkopf & Härtel about printing. In July, Müller did indeed go to Switzerland to visit van de Velde. Surely they not only enjoyed the fresh mountain air but also did business. After that, Müller continued on to see Titus Tobler, who lived on Lake Constance. A day after Müller's departure, van de Velde started correcting proofs again. He found that there were still so many mistakes and advised Müller to ask Leipzig to be extra careful. The map had to be out by October, when travellers left for Palestine.[119] Müller sent him the Jerusalem map, fresh off the press, and the final version of his Palestine map, but only so he could check the colouring. He urged a quick reply as he needed to have the hand-colouring done in time.[120]

The memoir was almost ready and Müller was busy preparing the next steps, which meant looking for potential reviewers and planning more Palestine maps with van de Velde. Karl von Raumer's Palestine map in the *Hand-Atlas*[121] needed to be completely reworked. Von Raumer required a cartographer and Müller thought van de Velde was perfect for the job. Yet von Raumer was old and van de Velde would have to submit to his scholarly point of view. Perthes also wanted to produce a smaller school map of Palestine and a wall map for schools. Müller actually wrote that von Raumer needed 'a draftsman to lend him a hand'. It is of little wonder that the proud and stubborn Dutch cartographer could not accept this offer, as became apparent further down the line.[122]

When receiving the corrected proofs from Paris, Müller saw that van de Velde had a right to complain. He immediately wrote to Leipzig and urged them to work more carefully. In the correspondence, there really is a letter by Breitkopf & Härtel promising a more thorough job and internal copy-editing before sending out the next proofs.[123] Over the summer, van de Velde changed his quarters in Switzerland and the mail was delivered to the wrong village. In the meantime, Leipzig was waiting for his work. The publishing date of 1 October could obviously not be upheld while the cartographer was enjoying walking tours in the Alps. He wrote to Müller how much more pleasant this was than sitting at his desk.[124]

In September 1858, van de Velde sent sheet 18 of 22 for the memoir back to Leipzig. It looked like things were drawing to a close, he wrote. He informed Perthes that he had recently been made a chevalier, a title to be added to his publications. On section 6 of the map, he asked Müller to please have the error corrected that named him a 'Late Lieut. Dutch R.E.' instead of 'Dutch R.N.' – the Dutch Royal Navy was not to be confused with the

Royal Engineers (interestingly enough, the finished map still bears this mistake). The Dutchman finally added a list of people and institutions to whom he wished to present the map as a gift as they had helped his cause greatly.[125] This included the publisher A.K. Johnston at Edinburgh, scholars and geographers who van de Velde owed for his work (Rev. Francis Thrupp, Carl Ritter and Heinrich Berghaus), military and government institutions who had probably supplied him with material (the U.S. Navy, the British Board of Ordnance and the Foreign Office) and European diplomats and monarchs (the Dutch ambassador at the Sublime Porte, the Ottoman sultan, Queen Victoria and the Dutch king). A last group was most likely sponsors in financial or other ways: 'Dr. C. Hingston, Plymouth, Rev. W. H. Carpendale, Leeds; and Mrs Wirgman for Mr Wirgman, Paris'. Unfortunately, none of these can be clearly identified.[126]

October came and the memoir was still not ready. At his Swiss summer retreat, van de Velde was forced to stay until the memoir was finished; there were two or three more sheets to correct. It was getting cold in the mountains. Since his sister and her family could not travel alone, they were forced to stay as well. Needless to say that van de Velde was not at all happy. He urged Müller to ensure he would not be disappointed again.

Then, in late October 1858, it was finally done. The Perthes correspondence contains a small slip of paper, the delivery bill of the complete proofs of the memoir from Breitkopf & Härtel to Justus Perthes. The first copies were shipped out in November.[127] From signing the contract to delivering the finished product, the production of the map and accompanying memoir had taken three years and four months. Considering that van de Velde had measured and travelled throughout the region in 1851–2, the total turnaround for publishing the geographical knowledge he added to Palestine research was over six years, although van de Velde tried to compensate by adding the latest research results to the map up to the very last moment. Although the map was only made available to the public in mid-November 1858, the publisher decided for 1858 as imprint. He did so most likely because the map took so long to produce that its information was running the risk of being outdated. While this seems a minor detail, it is telling for the timeliness paradigm that had taken hold in the geographical discourse of the nineteenth century. Geographical information not only needed to be scientifically generated and validated; in times of rapidly increasing knowledge it also needed to be as up to date as possible.

Yet the work had only just started. Immediately after the last corrections, the publisher had to spread the word and get the finished product to his customers. This is reflected in the many letters to and from potential reviewers as well as long lists of subscribers to whom the map and memoir had to be mailed. In the United Kingdom, Perthes's English business partner Williams & Norgate was in charge of sales and delivery.[128] Müller sent a copy of the map and memoir to van de Velde himself and informed him that copies would be shipped to subscribers the following day, 20 November 1858. Yet why make Prof. Berghaus a present of the map? Perthes saw no reason.[129]

To recover, van de Velde had relocated to the Côte d'Azur. The picturesque city of Nice with its mild climate and soft light was a popular resort for artists in the nineteenth century, and also for the English upper classes. Its main seaside promenade still bears the name Promenade des Anglais today. His copy of the map had failed to reach him at Paris, van de Velde wrote, could Müller send it to Nice? His health forced him to renounce any serious studies over the winter, so no more cartography for Perthes. Once recovered he would reconsider; but for the time being he had to refrain from work. For the rest, Müller knew the spirit of his contract with Bernhardt Perthes, and as long as Müller stayed head of the publishing house van de Velde would confide in him completely. If for health reasons, caused by his geographical travels in Palestine, he might be forced to eventually put this work into other hands (hands that Perthes deemed fit), the least Perthes could do was to share the profit, as his map would be the original and the source all following maps would be derived from. But he did not want to return to matters already cleared between them, and so he would let Müller do whatever his Christian heart advised him.[130]

Cartographic authorship and the limits of their contractual agreement were the subject of the letters to follow. The ensuing quarrel worsened the already fraught relationship between publisher and cartographer. Müller wrote how sorry he was that van de Velde could not currently do any cartographic work – he would have liked him to contribute his knowledge to more maps. Of course he would meticulously adhere to the paragraph of their contract that forbade him from making a discrete new map from van de Velde's. Yet with a clear conscience he would use it for other products of his enterprise, as any map publisher would, such as school atlases or for updating their resident cartographer Emil von Sydow's school wall maps. Müller just wanted to explicitly mention this, he wrote, to prevent further allegations which would hurt him so much more as he had pointed out how beneficial a wall or school map by van de Velde himself would be. A bit acridly he added that the map's production costs would surely not be covered by sales in the first years, judging from his experience, which made him feel most sorry for van de Velde.[131]

Van de Velde's explanation for sending a copy to Berghaus reveals an unexpected source of information: Heinrich Berghaus, van de Velde replied to Müller, had provided him with the results of Robinson's travels of 1838, so he wanted to reciprocate.[132] For the rest, van de Velde was surprised about the production costs. From his lithography album he had some experience with costs, and he would much regret to have ruined his health for this map if Müller was right. Supposing this was the case, it looked like Perthes covered the costs through other channels, namely using the map's data for their other atlases. In the end, it was the poor cartographer who had sacrificed years of work, considerable sums of money and his health for the profit of a publishing house, he wrote.[133] Müller's answer was short and clear: the production costs for map and memoir had been around 4,000 Thaler. He would be delighted if that was covered by sales in the next two years. Apart from that, he asked van de Velde to henceforth send letters regarding his map to the 'firm Justus Perthes'.[134] This separation of personal and business relations apparently made it easier for the two men to have arguments while maintaining personal relations. The cartographer's next letter went to 'Messieurs Justus Perthes', with an added 'Very friendly regards to Herr Muller [sic]' at the bottom.[135]

CRITICS AND COPYRIGHTS

While the two were falling out, the first reviews came in. In private, there had been favourable comments by Jakob Fallmerayer, Karl von Raumer[136] and Titus Tobler, Müller wrote early in 1859, adding that Tobler could not spare any criticism.[137] Criticism also came from the French geographer Emmanuel Guillaume Rey, who had discussed the depiction of the Hauran mountains as flawed at the Société de Géographie.[138] Then again, Müller reported that Alexander von Humboldt had written to him that 'among all recent endeavours [for the geography of the region], the atlas of the Holy Land by van de Velde is one of the most important'.[139] The reviews that appeared by and by in the United States, England, Scotland, France, Germany and the Netherlands unanimously praised the 'beautiful representation of the country that [...] supersedes all earlier publications of the kind', as the London *Quarterly Review* put it:

> Other good maps of Palestine have indeed been compiled during the last half century: among which we may particularly mention that of Berghaus, published in 1835, with an explanatory memoir; that of Zimmermann in 1850, which accompanied Ritter's volumes; and that of Kiepert in 1852.[140] But the map of Van de Velde far surpasses them all in fullness and accuracy, in clear delineation, and (no slight recommendation) in the cheerful and pleasing character of its colouring and shading. The author had great advantages for the task from his previous experience in the hydrographical surveys of the Dutch navy. But the construction of this map was almost the settled purpose of a life; and, undertaken as a labour of love, it was prosecuted and completed with a serious sense of religious duty.[141]

The beauty of the geographical knowledge visualised in the map was evidently as important as the quantity and quality of its data. The same compliment for clarity and style came from the German orientalists Heinrich Ewald[142] and Friedrich Arnold.[143] The French geographer Victor Adolphe Malte-Brun lauded the many details that could not be found in any other map and declared it the most complete and appealing map of the region.[144] As an aside, the publishing house not only sent the map to subscribers and reviewers, including luxury copies for noblemen, diplomats and sultans. Perthes obviously had an exchange system with institutions, too. For example, the British War Office thanked Perthes for the map and memoir and sent a number of their own maps in return.[145]

The map had been out for six months, yet van de Velde was not happy. He felt discouraged by the economic outlook Müller had presented him with and started blaming the publisher that, if the map was no success, this would be his fault.[146] Müller became angry. As publisher, his interest in the map's success was just as vital as the cartographer's, he wrote. Who said he did not do enough to enhance sales? Sending maps on commission to booksellers who were not associated with the Leipzig bookselling trade, as the Dutchman had suggested, was a sure loss; they would never be compensated. For that matter, France, but also Spain and Portugal, were the worst countries in the world for selling maps. Nevertheless Perthes would place ads in London and Paris once again, he added. For the rest, Müller asked van de Velde to be patient and refrain from offensive remarks.[147]

This rather harsh letter was probably delivered to Lausanne, where van de Velde had stayed earlier, so he got it much later. His next letter from July 1859 curiously enough speaks of the 'amiable message' he had just received. Quite evidently van de Velde wanted to make up. Half-jokingly he reported on his visit with 'chevalier' Karl von Bunsen[148] in Heidelberg, who had promised him recommendations to the King of Prussia and Alexander von Humboldt:

> The chevalier kept me busy [...] with a plethora of learned questions, of the sort that made my poor head spin; [...] there is nothing more boring in this world than too much *Gelehrsamkeit* [erudition; German in the French original]. Finally the chevalier invited me back to Heidelberg in September to confer with Mr Parthey[149] from Berlin and five other scholars in order to clarify questions from the Onomasticon [...]. I for my part am almost sure [...] this reunion of scholars will throw a dark night of obscurity on the Onomasticon. Tired of the *savants* in Heidelberg I took the express to Basle and Sunday in the early morning arrived at Neuchâtel [...], happy to take refuge in the beautiful Alps with their beautiful lakes for a while.[150]

In December 1859, Müller sent the first annual statement. The production had cost 4,112 Thaler 20 Groschen 3 Pfennig; earnings were 1,463 Thaler 12 Groschen 5 Pfennig. This left Perthes with a negative account of 2,649 Thaler 8 Groschen. Or, to put it another way, map and memoir had recouped more than a third of the production costs in one year (according to the contract, van de Velde would get his share only after production costs were covered). Müller thought the map might sell just as well in the following year, but there was no guarantee. As a map publisher he was used to waiting years until even his best products brought him profit, he said. What he could offer, as the result was so unlike van de Velde's expectations, was to buy the full copyright from him.[151] The Dutchman answered that he was not surprised, as the map was not yet known in the Protestant world. He was very content about the favourable reviews, all the more as he had done nothing to achieve them (seemingly unaware of what his publisher had done). After a few months of rest, van de Velde wrote, he felt strong enough to tackle Palestine geography once more. He had made arrangements to meet the 'savants' at Heidelberg in spring, and then, maybe, go to Palestine again:

> [T]he idea fascinates me, that as I am an established authority on Palestine now, it would not be in vain to go there a second time to correct my map in all the areas marked 'not examined' and to make a Biblical atlas. For this, I would need to be supported in advance by various religious societies from England, Switzerland, the Netherlands, and by the German scholarly community.[152]

If Perthes was interested, he would gladly continue the cooperation. In any case, they could not think of correcting the map before the first edition had sold out. Regarding the copyright transfer, he was well in favour of it, but asked for the conditions.[153] Müller outlined them as follows:

> By making your map my unlimited property, I understand that you are selling me the map, as it is now, for a fixed sum. I do not mean that I will keep it unchanged, although I will of course keep your name as its author. In fact I really want to buy the right to always keep the map up to date, which means to have it corrected, be it within or without my firm, and according to my wishes. I also want to do with it as I wish. For example, making a wall map or use a section for another purpose [...] – in short, I want to do with it whatever I please, except dishonourable things.[154]

He offered 1,200 Prussian Thaler for the map. The memoir, he wrote, would not be printed again, so, regardless of its value, he could not pay for it. It would take years for the production costs to be covered; with each new batch printed, more costs incurred. Yet he would consider the map copies that van de Velde had received so far as free of charge. Concerning a second trip, he was all for it; they could discuss that when van de Velde was in Germany. As requested, he sent a copy of map and memoir to the RGS. He would grant free copies in the future as well, but wished to fix a maximum number. Along with his letter, he sent a copy of Tobler's new book which contained a few things that van de Velde might not like, he added with slight irony.[155]

Their agreement on the copyright is another example of how both men kept business and private relations separate. They each wrote an official letter to and from the Perthes firm, then added a short private note. Yet a clash was inevitable. Again, van de Velde was not content, as he wrote to Müller in private. For him, the memoir was the most valuable outcome of his entire research; he found it hard to believe that Müller did not see this. Confidentially spoken, van de Velde had made 30,000 francs with his album, he said, with fifty more copies left. Work on the album had been nothing compared to the map and memoir. Yet, if Müller accepted the modest sum of 300 Thaler van de Velde suggested for the memoir, the deal would be sealed.[156] Müller's answer as the head of Justus Perthes reveals a central aspect of copyright in the nineteenth century: unlike the map, he said, everyone was free to use the memoir as they wished, without asking. Thus, from a commercial point of view, it was less valuable, whereas he saw its scientific value well enough. To end the debate, he agreed on the 300 Thaler but withdrew his offer of free copies. He asked for a simple 'yes' or 'no' for an answer. His private answer was, once more, short and clear: 'I believe I have done everything possible to accommodate your wishes. [...] I now rest my case whatever your answer will be and remain obediently yours, Adolf Müller.'[157]

At the end of January 1860, van de Velde accepted the purchase.[158] The most complicated issue turned out being how to transfer the money from Gotha to Lausanne and in which currency – if as 'billet pour Paris' or 'traite sur Genève' or 'Wechsel auf Frankfurt', as they finally agreed, to be issued in Floren (guilders).[159] Although they were still haggling about the number of maps to be subtracted from his remuneration, van de Velde acknowledged Perthes's generosity.[160]

In May 1860, they discussed further Palestine maps: Van de Velde thought of a folding canvas map for travellers, while Müller wanted a wall map for schools.[161] In the meantime, the publishers A. K. Johnston from Edinburgh enquired if Perthes would agree to a reduced version of the map to be used in the next edition of Johnston's atlas. They could offer an exchange of copyright, their new map of Scotland for van de Velde's map of Palestine. Müller refused:

> The intentions, which I have myself with this map, do not allow me to leave to another person its copyright. – I am, therefore, totally persuaded that you will make no use of the map in such a way, that my claims on it are offended by your publication.[162]

A school wall map, Perthes wrote, would have to be done in the design their resident cartographer Emil von Sydow had established: the terrain in big, dramatic ranges, more painted than drawn really, printed in colour print and lithographed. Considering van de Velde's talent, why not align with von Raumer for this purpose? Perthes wanted to do two school wall maps: modern Palestine and the lands of the scriptures, as they had understood that displaying both in one was impossible.[163] Again, van de Velde declined, as he feared not getting along with von Raumer, he said. The idea overall seemed too stressful to him, to which Perthes agreed.[164]

Due to the political situation, van de Velde postponed his plans to go to the Holy Land. He described the situation (violent conflicts between Druze and Maronite Christians in Lebanon and Syria with many dead) in a remarkable mix of Protestant millennialism and mercantile interests:

> This unfortunate country has now entered the phase of its final battle, a battle that will last longer than we think. Today it is Muslims and Christians who are fighting each

> other. Soon it will be Catholics, Greek Orthodox and European Protestants who will declare war on each other who over owns the country. Yet the Word of God promises it to none of them but to his own people, the Jews. Whatever the outcome, the map will surely be in high demand.[165]

In 1861, van de Velde left Lausanne for good, as he wrote to Gotha in June: 'I say thank God that I left this backwater town!' In autumn, he would be staying in Geneva with Henri Dunant (the later co-founder of the Red Cross), then in the Swiss Alps in the summer with his sister and her little girls – they were asking if Müller would not join them.[166] Yet, this time around, Müller preferred spending his summer holiday elsewhere.

UPDATING THE MAP: SECOND TRIP AND SECOND EDITION

A cornerstone to the success of Justus Perthes publishers was the fact that they continuously updated their cartographic products. Advanced by growing European influence as well as research and travel activities in the region, the *Map of the Holy Land* was to be thoroughly revised. Van de Velde combined this with his humanitarian work which was becoming more and more important. He travelled, not only as a cartographer who wanted to amend his map, but also as an emissary of the British and Foreign Syrian Asylum Committee to investigate the state of the Christian population in Lebanon and Syria after a massacre in 1860.[167]

As can be seen in the correspondence, updating the map was a walk in the park compared to the tumultuous years before it first came out. In October 1861, van de Velde swiftly decided to go on a second trip to the Holy Land in November. If Perthes paid his train ticket, he would gladly visit Gotha to see the latest publications before he left, he wrote (and Perthes agreed). His plan was to go first to Paris, then to London, both to consult new material, and finally to Gotha.[168]

The pace of events quickened; updating the map for a second edition was well underway. Van de Velde travelled to Paris and London, Haarlem and Gotha, Erlangen and Geneva within only six weeks, then finally to Marseilles where he boarded a ship to cross the Mediterranean.[169] He reached Beirut on 16 December 1861. First trapped in bad weather, he ultimately travelled through the Galilee and back and forth through the country, measuring, via Nablus and Ramallah to Jerusalem. He even wrote to Gotha from Jerusalem saying that the Spittler Mission[170] general store would be an ideal point of sales for the map and that Perthes should send maps there.[171]

Fast-forward to the published second edition, we can see the effect of his second trip on the map. Section 5 for example shows the country between the Yezreel Valley and Jerusalem, as well as the Mediterranean Sea and the river Jordan. In the first edition, van de Velde had marked areas he had not been able to survey as 'not examined', to make his own work sufficiently transparent and as a call to action for further explorations.[172] When aligning his surveying route with third-party sources he later named in his second memoir,[173] we can retrace how the map was updated in detail. In 1851–2, he had for instance not been able to measure between today's Nablus and Qafr Qasim (Kefr Kasîm) due to extreme weather.[174] In 1862, his travel route in this area covered two 'not examined' areas of the first map that consequently vanished in the second edition.[175]

From Jerusalem, van de Velde made a side-trip to Hebron, then from Jerusalem on to Jaffa, from where he went to Beirut by ship on 6 April 1862. He embarked for Marseilles on 13 April, landing on French soil again on 2 May. Without delay, he proceeded to Paris to examine the new French ordnance map of Syria, then to London to consult the latest Admiralty maps. In June 1862, he was back in Geneva after eight months. Already from London, he was summarising his findings for Perthes:

> I gathered some new observations and made a few new discoveries of antique places, most of all between Hebron, Beit-Jibrin et Beit-Jala; and in the Galilee between Tyr, Alma and Bukeia and Nazareth. Other geographic contributions have been given to me by Dr Rosen in Jerusalem and the missionary Sandreczki.
>
> I have just examined the new French ordnance map between Sidon, Damascus and Acre in Paris and some other new maps of the coast of Syria at the English Admiralty. There are also the observations of Wetzstein in the Haurân. All that forms the basis of a new map of Palestine, or to be more precise a new edition of my map.[176]

If Perthes was inclined to produce this new edition, van de Velde kindly asked for the conditions, closing with cordial greetings to everyone in Gotha, the Müller, Besser and Perthes families.[177] Müller asked how much of the map had to be changed as he needed to calculate costs. How many sections would have to be re-engraved? Could van de Velde send a sketch of his corrections? The Dutchman had to keep in mind, he wrote, that 200 Thaler of the production costs still remained to be covered and the map had not been in high demand lately. Wetzstein had by the way announced a more thorough publication to follow, so this section of the map would run the danger of being outdated if they did not wait for his results.[178]

Once more, a debate about costs and the value of cartographic work ensued. Van de Velde again brought forward his Protestant values of trust (between him and the publisher), honesty (of both parties toward each other) and Christian morals ('cœur chrétien'). He also claimed that it was not his fault the map had not sold well; he had transferred his copyright to Perthes. In short, he insisted upon the non-commercial value of his work, whereas Müller had to make economic decisions. Other than that, van de Velde had asked his French publisher, Lemercier, about the costs of lithographing a second edition in Paris and sent Müller the estimates.[179] Yet Müller could not decide on having the map recast for lithography:

> You will remember that a while ago you were convinced that your map, engraved in copperplate, would provide knowledge of the Holy Land for a long time. Why should I now throw the copperplates away and have a new map lithographed [...]?

It was clear to Müller, so he said, that at least two sections had to be engraved anew. For every new section he could offer 150 Thaler, for minor corrections he would pay according to the extent of changes. Of course, the copyright would remain his.[180]

The Dutchman accepted but not without complaining that his publisher did not understand his best intentions. He had neither the time nor the energy to waste on another memoir, but would write a few lines of explanation, he said. He also agreed to cede the copyright to Perthes, yet asked to be supplied with all maps and materials he needed. Hermann Berghaus, the nephew of Heinrich Berghaus and resident cartographer at Perthes publishers, had kindly compiled a list of the latest publications for him when he was in Gotha in 1861; van de Velde now asked for an update (see Fig. 2.6). He also asked for a copy of the map to sketch the corrections on and a print 'without mountains' (so we know that the map was printed from two copperplates, one for line engravings such as borders, rivers, towns, and one for the mountain relief). His would be staying with his sister, as he was going to the Netherlands for two months to hold public lectures on the situation in Syria.[181]

Obviously Justus Perthes was not only a publishing house, but a complete research institute, with library services and research reviews compiled on demand: all the latest publications, maps and other materials were swiftly sent to Haarlem.[182] The process of assembling and verifying geographical knowledge (by retracing its production) for the second edition was very similar to the map-making in the first place. It also suffered from some of the same problems that delayed the initial publication, namely waiting for third-party data. Nevertheless updating the map, including a second surveying trip, took only half as long as making it in the first place.

While van de Velde was reworking the map in winter 1862, John Washington of the British Admiralty informed him that the survey of the Mediterranean coast, directed by Commander Arthur Lukis Mansell, would only be completed in the following spring.[183] Mansell had already headed the Navy survey of the Bay of Suez in 1856. From 1860 to 1863 he was commander on HMS *Firefly* and made several surveying trips to the 'coast of Syria', including Palestine.[184] The survey ultimately linked the longitudes measured to verified positions in Malta, thus connecting Syria to European grids.[185]

The raw data of the astronomical observations by the German engineer Richard Doergens, however, were out of reach for van de Velde.[186] Müller urged the Dutchman to at least write to Heinrich Kiepert, who was supposed to make a detailed map from Doergens's data. Perthes would not get anything from Kiepert because they were competitors. Regarding the area north of Damascus, they had been unable to find any information and could not help him there.[187] Yet Kiepert, van de Velde wrote back, was just as much his competitor as Perthes's, so writing to him was useless.[188]

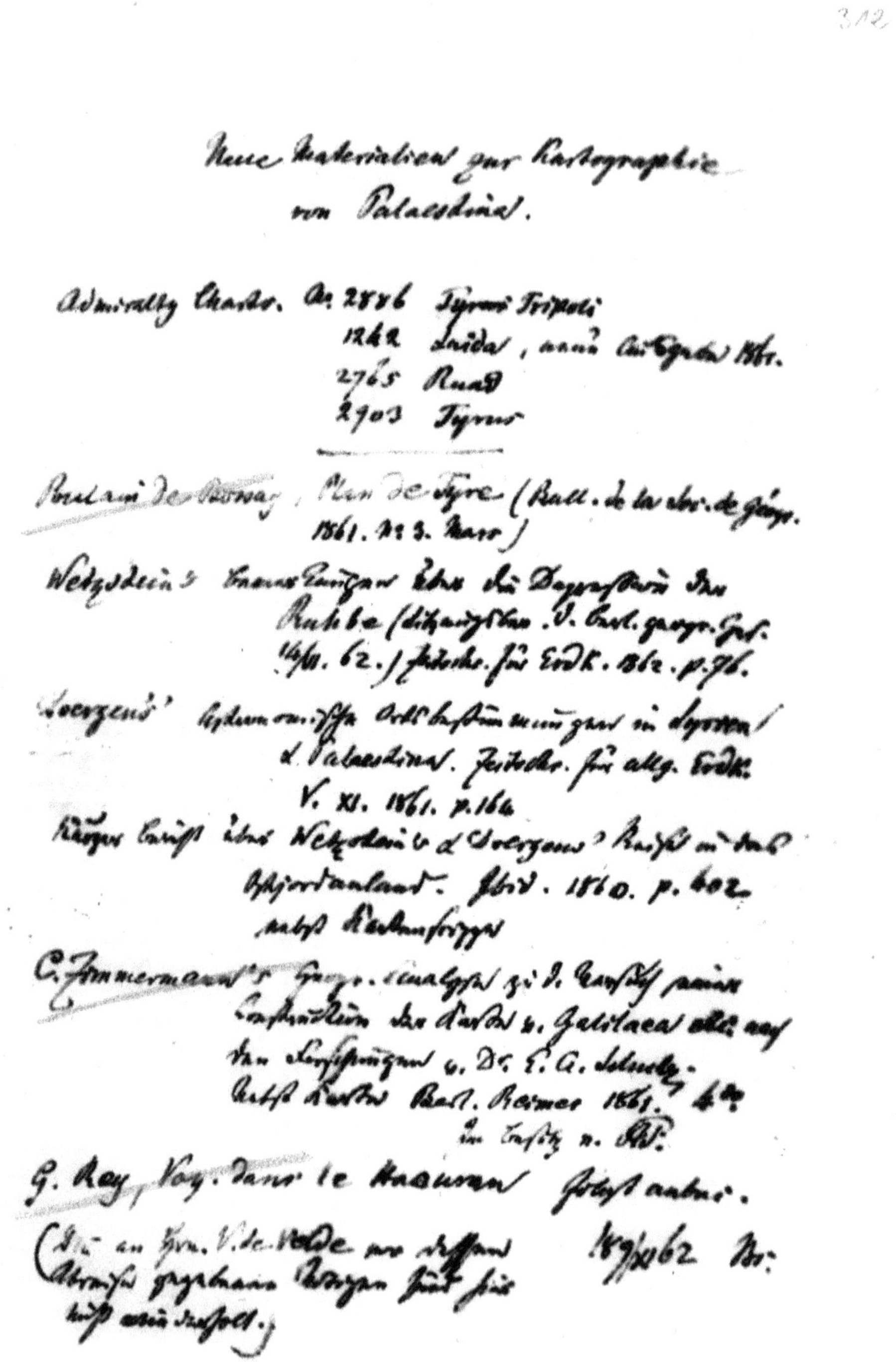
Neue Materialien zur Kartographie
von Palaestina.

Figure 2.6 'Neue Materialien zur Kartographie von Palaestina' (New Materials on the Cartography of Palestine), manuscript by Hermann Berghaus, November 1862; copy on transparent paper
Source: FBG SPA MFV 9 f. 312

All winter long, van de Velde worked on the map. In April 1863 he became ill again. He also received another letter from Mansell, which was so promising that he had to wait for Mansell's results by all means. Mansell had just travelled the region with the *Map of the Holy Land* in hand, finding it so exact that he decided to grant van de Velde access to all survey maps and measurements, triangular and astronomical, before publication.[189] Van de Velde enthusiastically wrote to his publisher that he would 'never[!] have to update the map again' once he had included Mansell's data. At least Doergens's results might be published while they waited for the survey.[190] Kiepert, as Müller quoted from a letter received from him, was waiting for Doergens's contributions as well, so for the time being nobody had them.[191]

Van de Velde spent June 1863 in Bad Kissingen for his health, then the rest of the summer in the Alps again. While waiting for Mansell's results, he only resumed work on the map in December. He had found the French map of Lebanon[192] full of mistakes and corrected it with astronomic positions supplied by Mansell. The work was progressing well and he hoped to complete the map in early spring. Once more he asked Perthes to send him the latest material on Palestine. By the way, he had found Doergens's measurements worthless when comparing them to Mansell's, which came as a relief as they had differed from his own.[193] Müller dutifully sent another list of latest publications; he suggested publishing a short memoir in their journal (*Mittheilungen*).[194] That December in Geneva, van de Velde together with a group of other delegates, also founded the organisation that was to become the International Red Cross.

When van de Velde finally obtained Mansell's results in March 1864 he was overjoyed. Mansell sending him his half-finished map, months before its publication, was the most generous act he had seen in a long time, he wrote happily. What was more, Mr Hull, a member of the surveying team, had added a letter with 'the greatest praise that my map and memoir could have received'. This, plus the favourable news he had received from the RGS, would grant his map great success in England.[195] Only a few days later, he sent the finished map to Gotha with the corrections done on transparent tracing paper. The mountains had contour lines now, according to the latest cartographic fashion (a feature the publisher ultimately did not add to the map). So as not to burden the map any further, which was already very much charged with information, they should not colour his old travel route again. Concerning the political divisions, he had not been able to receive additional information, so it was useless to indicate them. He also sent the complete text for the new memoir, although they agreed it should only appear shortly before the map was published. Yet he wanted to put the books aside and feel that his work was finally done.[196] As usual, he added a personal note for Müller, thanking him in warm and amiable words, but complaining about the hardships of his task again:

> What can I tell you about my work? I would have broken down had it continued one more month. It seems that my health does not tolerate this kind of map work anymore.

> No, it's done! And with utmost precision and stern discipline. With it, you now have again the best and most current [map] there is on Palestine (west of the Jordan).

A war in the region was sure to come, he added, so the engravers had better finish the map soon. Speaking of war, he was about to leave on a mission to the Second Schleswig War theatre as a delegate of the newly founded Geneva Committee.[197] His cartographic work had to wait. Only after his safe return from the Dybbøl fort and another summer in the Alps, could work on the map continue.

The following letter was written to Adolf Müller personally and speaks about the crucial transition from sketch map to engraving. Since it was unclear how the neutral emissaries would be treated by the warring parties, the staunch Protestant had obviously made special arrangements. Before his departure for war, as he put it, he wrote to Gotha:

> I realised from the beginning [...] that the engravers should have used my sketches on tracing paper [but] [t]here were corrections after corrections [...] so that my sketch map, although good in many places, has become so unintelligible in many other parts that I was afraid of tossing the engravers into a ocean of confusion [...]. But if it is convenient for you that I send my sketch drawings, the same will be at your disposal [...]. Yet I can only send them to you when I come back to Geneva healthy and alive in a few weeks, since I packed all of my things away like someone who is going to war [...]. But the engraver should, I think, make the drawings on tracing paper himself, first on the new sheet, then fit them on top of the old ones. That is how it has always been done, even for the previous maps he made much more complicated drawings on tracing paper himself. [...] Only explain to the engraver that he should first make a copy of all the new sheets on tracing paper; then, when he puts these on top of the old ones, not a single ambiguity should remain for him. [...] I barely need to say that I wish to help especially the German wounded in Denmark. Can I do anything for you while I am there? Do you have an agent there?[198]

Both van de Velde and Louis Appia, the second emissary of the Geneva Committee at the Dybbøl fort, returned unscathed. Eventually, in autumn and winter 1864, the revised map sheets, proofs and corrections were sent back and forth between Geneva and Gotha. The work progressed quickly and the Dutchman planned to sell 100 copies of the map using his own account, for which Perthes had granted him the bookseller's trade discount. From his lithography album, he wrote, van de Velde had learned that selling through personal connections was more profitable than through official agents. He had recently established good relations with Grand Duke Constantine,[199] had good relations with the RGS and the British Navy, so he was confident that selling the map through his networks would work in his favour.[200] Van de Velde also planned to make a biblical map for the French market, which was very different from his Gotha map, and asked for Perthes's permission.[201] He also needed to know when the map was to be published and if it would cost 30 francs like the first edition.[202]

Among other things, the engravers had difficulties connecting the mountain relief of the map's new sections 2 and 4 to the old one of the first map.[203] Only after proofs and sketches of all eight sections had been sent back and forth with van de Velde having to personally sketch the connection of the relief on the proofs himself, and after another four to five weeks of delays in engraving, the cartographer gave the 'bon à tirer' from Paris in April 1865. On 1 May 1865, Müller was ready to print:

> The German edition of the map will be done in such a way that I have the sections transferred from copper to stone and the English text changed to German on the stone. Maybe I will have the changes done in copper already. The political colourings will be omitted; instead I will introduce circa 9 altitude levels in full colour tones and the water blue in the lithography print. The price for the German edition will be 4 Thaler. To the readers of the *Mittheilungen* who get a sample sheet, I will offer it for 2⅔ Thaler even [...]. Maybe with this cheaper price, I can reinforce the map in Germany, which would be very desirable.[204]

Twenty luxury copies to be distributed among European monarchs by van de Velde, who had set high hopes on gaining subscriptions this way, were already in print. The price would remain the same, Müller added: 6 Thaler for the map in ordinary blue cardboard cover and 7 Thaler for the same in calico. He would grant van de Velde 33.3% discount on the retail price, plus the memoir free of charge for 100 copies to sell using his own account.[205] On 8 May 1865, van de Velde received his fee of 718 Thaler (including 68 for the memoir), or 2,638.90 francs, which makes it plausible that more than the two sections originally planned had to be re-engraved.[206]

Alarming news came from London: from a secretary of the Religious Tract Society, van de Velde had heard that an ordnance survey of Palestine was underway and the maps were to be published by the RGS. 'This would naturally become the Standard map for England', he quoted his (unnamed) friend. Van de Velde feared this would ruin the chances of his map.[207] As we know today, it took years before a survey of Palestine became reality. The earliest activity of this kind was the Jerusalem survey carried out by a party of Royal Engineers under Captain Charles Wilson from September 1864 to July 1865 on behalf of the Jerusalem Water Relief Fund.[208] While it is likely that van de Velde's London friend had heard about that, it is not impossible that the survey was initially supposed to cover a larger area. Upon van de Velde's enquiry, RGS assistant secretary Henry Walter Bates reassured him concerning an eventual map, and offered the society's backing – yet he interestingly spoke of an 'Ordnance Survey of Palestine', which would be the earliest mention of a survey of this scope:

> They [the RGS] look forward to [the map's] approaching completion [...] & if you send an early proof copy – will forward their order for the number they require in time for you to use the name of the Society as favouring your work.
>
> The 'Ordnance Survey of Palestine' to which you allude is undertaken by Sir Henry James; this Society [lending] one half of the expense. I do not yet know whether the Map resulting from the Survey will be published in our 'Journal'.[209]

Although the map was finished and on its way to Saint Petersburg, London, The Hague and Constantinople, the Dutchman was not satisfied. He repeatedly argued with Müller about his discount, about freight costs and extra free copies. When Perthes announced the German map at a reduced price, he got angry because he feared nobody would buy the English map. Indeed, Perthes had lowered the price for the German version once more, the retail price being 2½ Thaler or 1⅔ Thaler with the coupon printed in the *Mittheilungen* (see Fig. 2.7).[210]

When the RGS ordered only six copies, van de Velde was so enraged he withdrew his order for 100 copies to sell using his own account.[211] Müller accepted the cancellation but saw no reason for English speakers to buy a German map.[212] And, indeed, the international phase of Biblical Geography was approaching its end. While in 1854, when Bernhardt Perthes and van de Velde first started negotiating, it was perfectly possible to make an English-language map for the German market. However, ten years later it seemed utterly unthinkable to sell a German-language map to an English-speaking audience. Also the German audience would not put up with languages other than their own anymore, as we will see.

Finally van de Velde thought his idea of sending complimentary copies to important people was bearing fruit: the Prussian consul in Jerusalem Georg Rosen had written favourably to him about his map; in this case he would not profit in the least from the increase in sales.[213] However, Müller dashed his hopes in his answer:

> Concerning your great expectations produced by the letter from my personal friend Dr Rosen, I can calmly inform you that Mr Spittler from Basle obtained 13 copies of your map in 1862, of which a considerable number still warmed the shelves when Mr Besser was in Jerusalem this year. Mr Spittler would like to get rid of them again as the customers keep asking for a French map that he is forced to keep in stock. Until today, he has not ordered the new edition and will hardly order a greater number in the future as he has not been able to sell 13 copies in 3 to 4 years. I hope this information will put your mind at ease about the 'least profit on my side'.[214]

In April 1866, van de Velde paid for six copies of his map that he had ordered from The Hague earlier. It was his last documented contact with the publishing house.[215]

Based on the sales figures in Justus Perthes's accounting books, we can actually determine that the map was not a commercial failure, although Müller was right about its long-term perspective: 570 copies of the map and 454 of the memoir sold in the year after it came out;

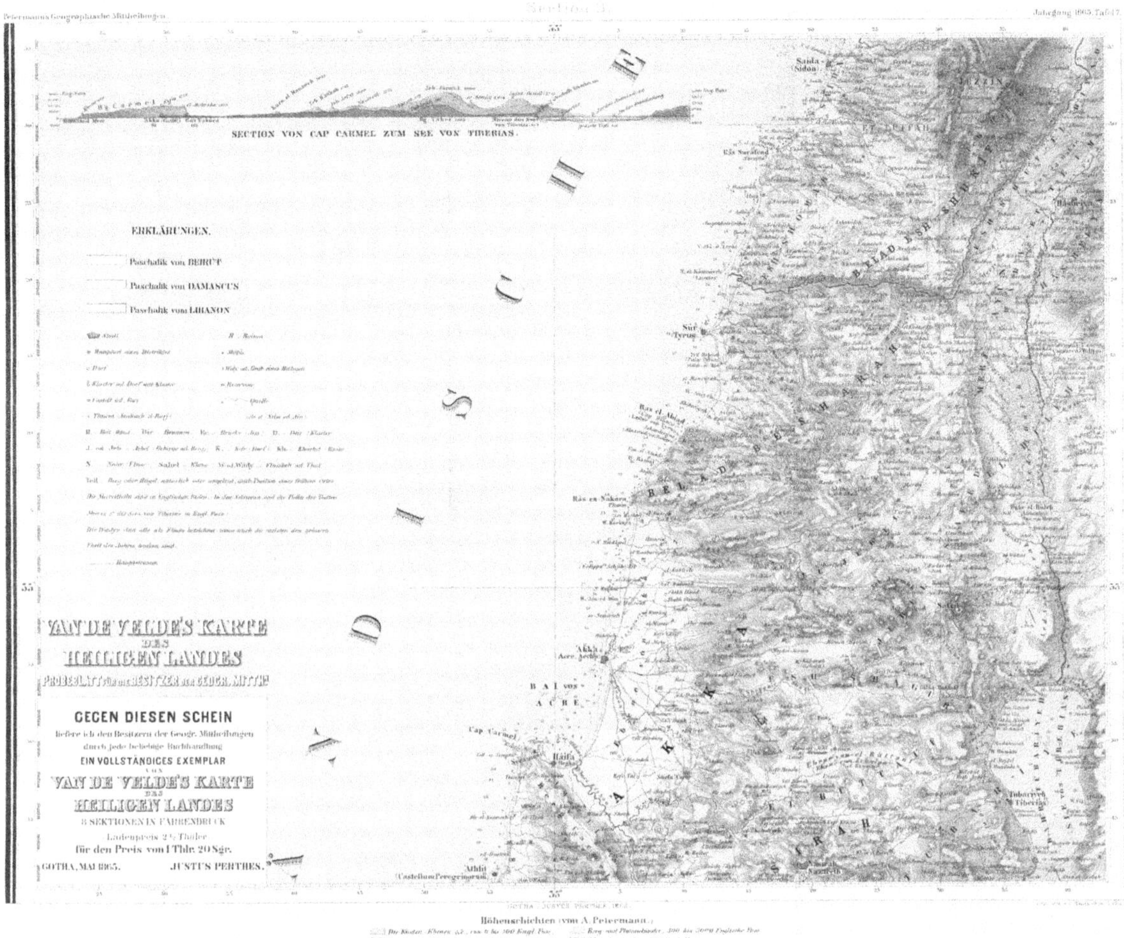

Figure 2.7 C.W.M. van de Velde, 'Karte des Heiligen Landes' (Map of the Holy Land, coupon from the journal *Mittheilungen*) (Gotha, 1865)
Source: FBG SPA 4° 100, vol. 11, plate 7

another 337 maps and 191 memoirs were sold in the following two years. After that, an average of about 50 maps and 20 memoirs sold yearly until the second edition was published in 1865. Altogether a total of 1,200 copies sold of the first edition; 1,000 had initially been planned in 1855.[216] Van de Velde was right about the advantage of the cheaper German version – almost 3,000 copies were sold by 1877 and only 327 of the second English edition. All in all, until the map was phased out of the catalogue twenty years after its production, Justus Perthes having sold a total of 4,409 copies of both editions. For comparison, the Jerusalem map by van de Velde and Tobler sold 612 times in the same period.[217]

Last but not least: when Titus Tobler returned from Palestine in December 1865, he wrote to Perthes: 'In the Spittler shop, the latest edition of van de Velde's map is displayed for sale.'[218]

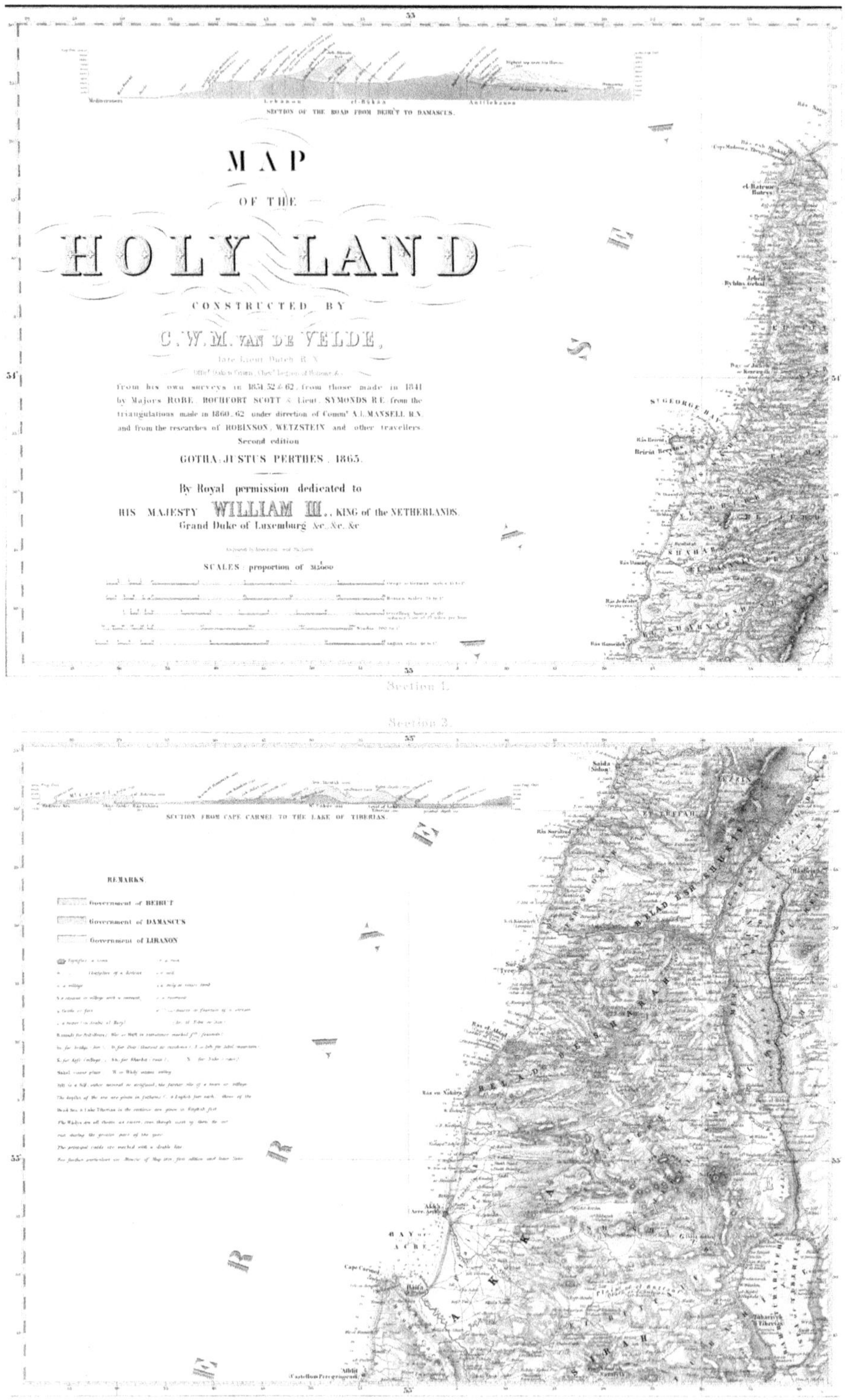

Figure 2.8 C.W.M. van de Velde, *Map of the Holy Land*, 2nd edn (Gotha, 1865), sections 1–8
Source: FBG SPA 2° 315

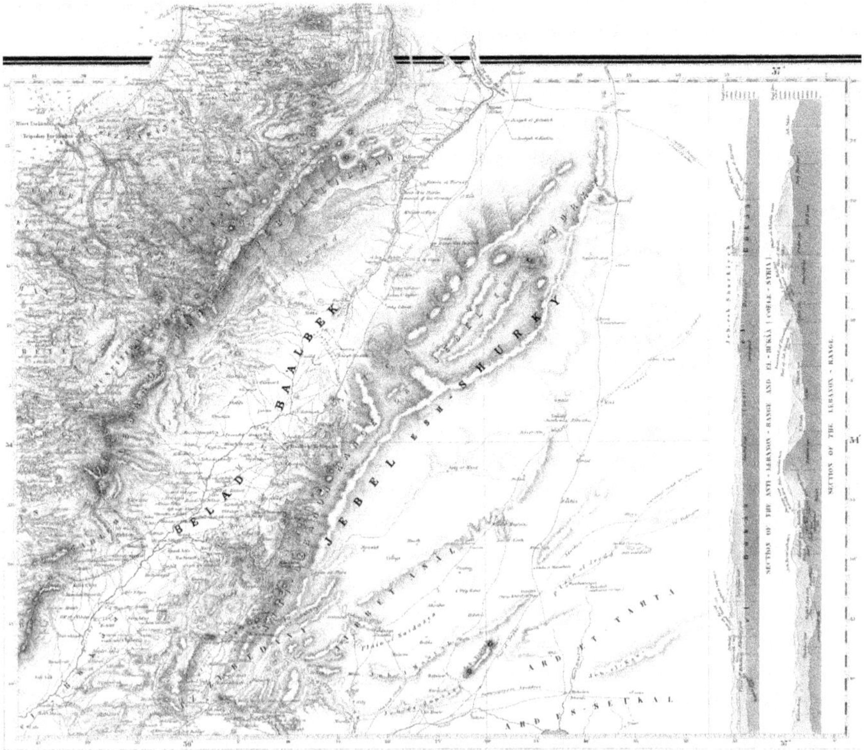

Section 2.

Section 3.

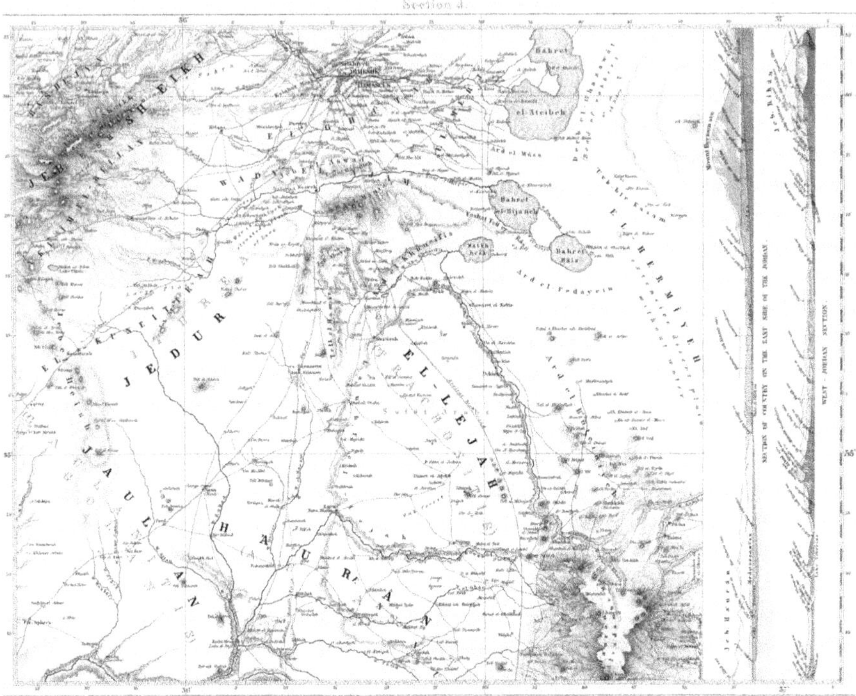

Section 5.

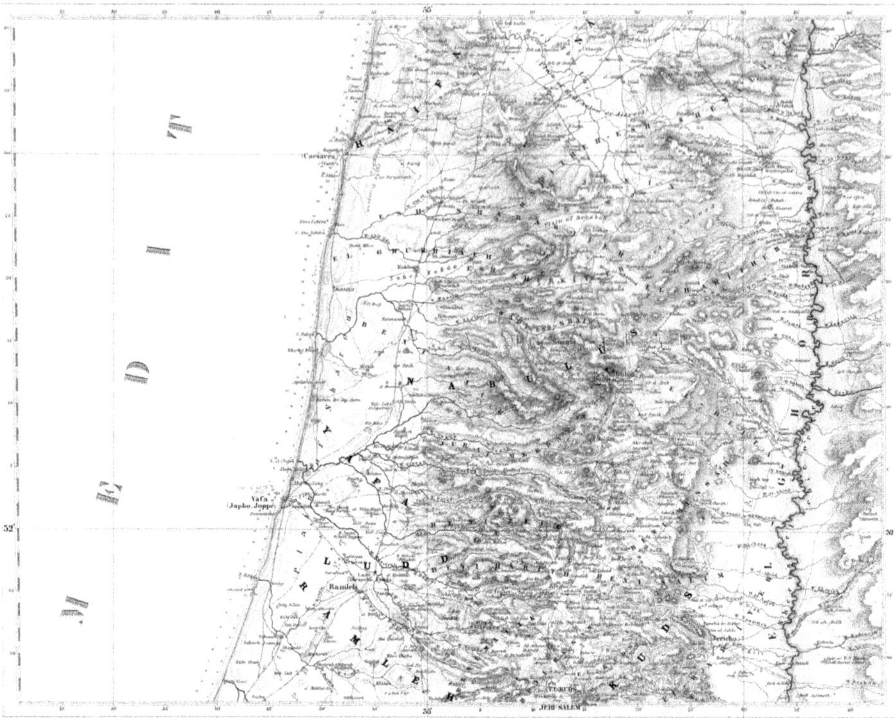

Section 7.

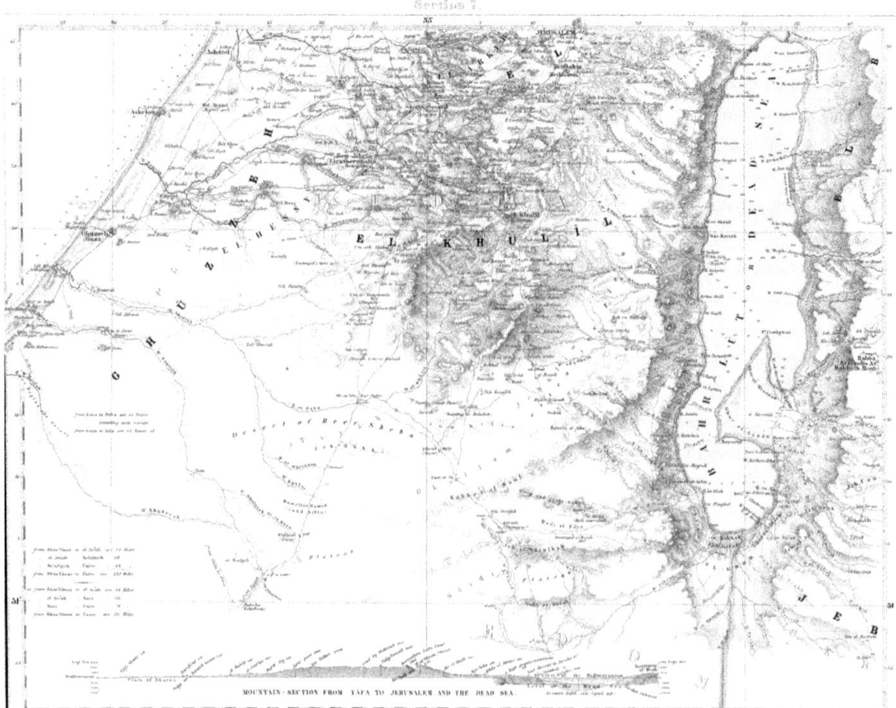

Section 6.

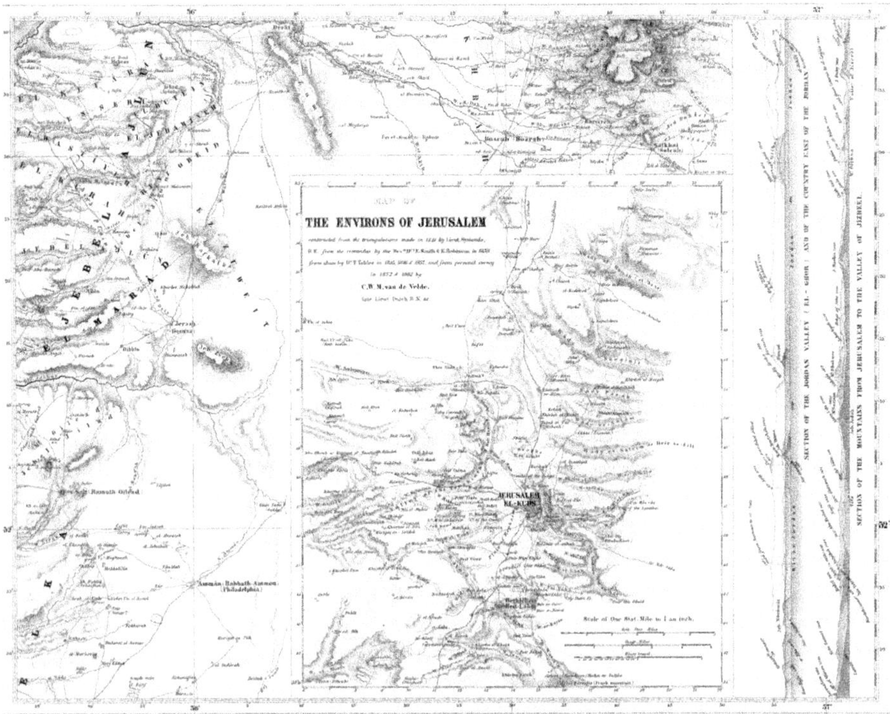

Section 8.

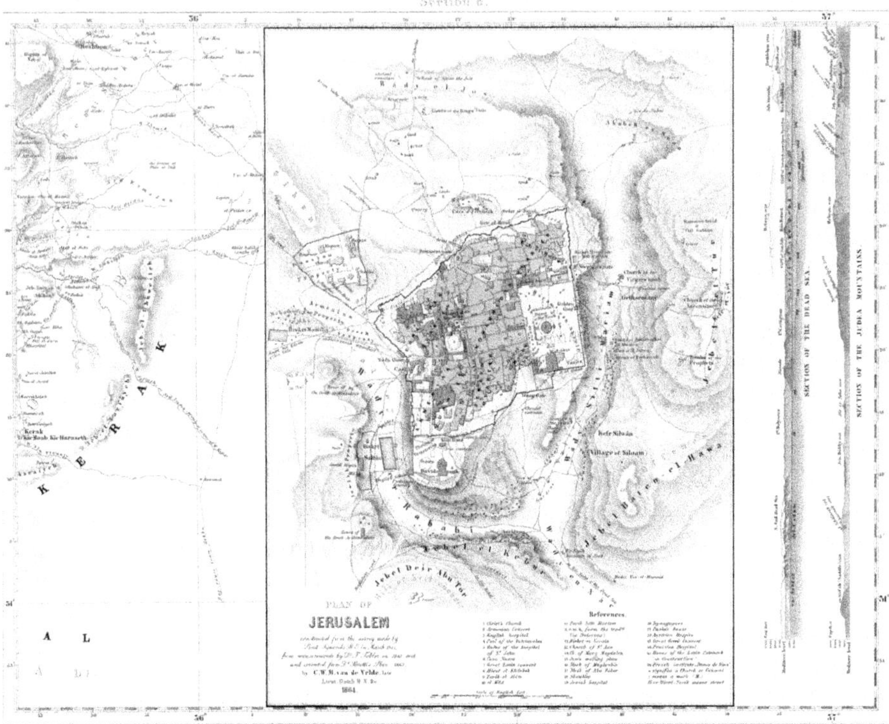

THE MAP'S AFTERLIFE IN THE PUBLISHING HOUSE

When the *Map of the Holy Land* had come out in 1858, German scholars and laypeople had unanimously praised its significance, yet criticised it for being in English and rather expensive. For a second edition, the cries for a translation into German rang louder. August Petermann, who had not been involved in making the first edition, consulted Friedrich August Arnold, Associate Professor of Oriental Languages and Literature at the University of Halle/Saale. Petermann's correspondence with Arnold give us a picture of the conflicting aspects of Holy Land cartography. The orientalist Arnold was concerned with the scientifically correct transcription of Arabic place names which were different in English thus suggesting a 'false' pronunciation for the German readers. Petermann, in contrast, was forced to keep the costs of reworking the engravings low. He pleaded for a pragmatic approach; they could not wait for the experts to agree on philologically correct spellings, he said. Actually, he wrote to Arnold saying that the characters on a map were:

> only an index for the maps, the illustrations of the earth's surface, and since they strain and weigh down the map so much, in many cases covering and pushing the actual geographical elements into the background, [they are] virtually an evil if at all necessary.[219]

When the map was translated into German, Perthes also furnished it with August Petermann's new system of colour representation of relief, thereby visually aligning it to the map style developed in-house. The German second edition was seamlessly integrated into the further map production of the publishing house, mainly as the basis for the following edition of *Stielers Hand-Atlas*, and just as seamlessly incorporated into their information processing and distribution programme, whose centre became their journal. The magazine granted its readers a discount, the memoir to the second edition and the cartographer's account of the map's making, including new geographical positions and other important data published there. Journals, as we can see, had become the new media of the day for a map memoir.

The Palestine cartography of Perthes publishers had gotten a whole new foundation from the map by van de Velde. As we have seen, they were so impressed by his skills that Bernhardt Perthes offered him a job, which van de Velde rejected on the grounds of poor health.[220] After the untimely death of the publisher, the managing director Adolf Müller renewed this offer, this time for the Palestine map in *Stielers Hand-Atlas*.[221] When the Dutchman declined once more, the slightly annoyed Müller had him know that he would, of course, 'like every other map publisher', allow the map to be used for further products in the house such as Karl von Raumer for the *Stieler* and Emil von Sydow for wall and school maps.[222] Von Raumer, who had provided the Palestine map for the *Stieler* after 1854, eventually died of old age in 1865. The Palestine cartography at Perthes was henceforth done by August Petermann himself, as Bruno Schelhaas describes in detail in Chapter 3 this volume, as well as by Theodor Menke with Friedrich von Stülpnagel drawing the maps. So for example Menke designed a stand-alone Bible atlas in the 1860s using Spruner's *Historisch-Geographischer Atlas,* van de Velde's map and the maps from the *Stieler.*[223]

MAP MAKING AS KNOWLEDGE PRODUCTION

As already discussed, in the cartography of Palestine, the *Map of the Holy Land* was part of a transition which had happened earlier in other sciences: the armchair scholar, who was thought to have a complete overview of all existing knowledge and was thus able to gather, evaluate and spread new insights from his chambers, lost his exclusive scholarly authority. Instead, the new paradigm called for knowledge gained by observation in the field. Alexander von Humboldt's South American travels were the prime example for this paradigm shift. The new explorer-scientists were experts who knew the area first-hand. This also gave rise to an antagonism between knowledge generated via fieldwork and the armchair scholar's study knowledge, as well as the respective types of scholars involved[224] – which is very palpable in van de Velde's comments about his 'savant' colleagues Titus Tobler, Georg von Raumer or Karl von Bunsen. Likewise, Tobler described van de Velde as someone who did not like to look things up in books, did not speak the right languages and wrote too sentimentally.[225] It is not surprising that van de Velde declined all offers to become a full-time cartographer with Perthes. His comprehensive approach to the Holy Land, as a devout Protestant and trained cartographer who also excelled in landscape painting and Biblical Geography, was becoming outmoded very fast. The experts on the various subjects, be it transliteration of Arabic, ancient place names of the Onomasticon or Biblical Geography, criticised him for not being up to date with their emerging scientific disciplines. While he remained unmatched in measuring and mapping, and was uniformly lauded for the beauty of his maps, he did not want to be confined to cartography alone. He refused to become a specialist at a time when the academic field was specialising, its discourse being shaped more and more by expert cultures. In this process of professionalisation, van de Velde had become somewhat of a fossil. After 1865, he consequently abandoned cartography altogether and henceforth devoted his time to art and humanitarian matters.

Yolande Jones, in her 1973 essay on the British Royal Engineers hitherto overlooked early triangulations of 1840–1, placed the *Map of the Holy Land* in the succession of modern Palestine maps based on triangulations, but declared it lacked originality:

> Van de Velde's map [...] is commonly held to be the first to mark the transition from compilation to a more thorough original survey. This is a misleading concept as Van de Velde probably made use of more sources of information than any other person who

> compiled a map of the Holy Land. Furthermore, it never was his intention to make a 'complete triangular survey' but 'a survey of a more superficial character, and yet satisfactory for the present times'. [...] Even so, the map which he produced was remarkable in one respect; much of it was founded on the basis of theodolite triangulations which had been carried out almost eighteen years previously. [...] It is clear that the true significance of Van de Velde's map lay in the material it incorporated.[226]

Apart from the fact that discussing a map along criteria such as significance, accuracy or originality is a completely different debate,[227] Jones's verdict somewhat blurs the point. Until the realisation of a full land survey of Palestine, which was difficult at van de Velde's time, both map compilation in the office and instrumental surveys in the field (when possible) were essential to cartography. This was as true for Palestine as for other regions mapped in the colonial nineteenth century.[228] So as well as asking to what extent the *Map of the Holy Land* used previous material (and to what extent it was original), we should ask *how* van de Velde used that third party data. For this, the 1840–1 triangulations by Lieutenant Symonds and others are an excellent example.

Although the Royal Engineers were technically well-equipped, their results were not without flaws. Due to fever and other hardships, Symonds had often operated the theodolite alone. He also miscalculated the heights, which resulted in a great confusion about the level of the Sea of Galilee.[229] Moreover, the Royal Engineers's documentation was not sufficient to reproduce their results.[230] Lastly, the group made a small but significant mistake in Jerusalem: They misinterpreted the outlines of the Haram, resulting in a faulty depiction of its western boundary. On account of this, Edward Robinson[231] and also August Petermann, as Bruno Schelhaas has shown,[232] considered the survey untrustworthy. Van de Velde, however, had been able to consult the original material at the London Admiralty and used its raw data to construct his own map, connecting his results to their triangulations. Only after critically compiling all existing previous geographical and cartographical knowledge as well as retracing the process of its construction, he ultimately combined it with his own measurements in the field.

Furthermore, van de Velde's excellent connections to military, religious, scholarly and political circles throughout Europe enabled him to gain access to sources that were not publicly available and, in some cases, would not have been published elsewhere. Yet he never used them without a critical assessment, perhaps with the one exception of the Royal Navy coastal survey under Mansell.[233]

In his pioneering study on the European rediscovery of the Holy Land, Yehoshua Ben-Arieh stated that 'Ritter's writing and van de Velde's map together may be considered the crowning achievements of fifty years exploration and research in the geography of Palestine.'[234] And, really, we can say that van de Velde's map represented the complete body of Western geographical knowledge on Palestine, critically compiled, then verified on site and completed by his own measurements and his own (physical) experience, including the hardships needed to make a scientific map, while Carl Ritter (who died in 1859 aged 80) had been the prime example of the armchair geographer's superior scope of knowledge. Two generations later, van de Velde would collect all available knowledge and resources with the practical eye of the experienced cartographer. He thus combined the best of both worlds, the office and the field, and verified it with his specific practical skills. Since he was not trained in the subject areas that were becoming more and more important in Palestine research, namely theology and oriental languages, he could not compete with the new experts in this field, the most distinguished of whom would be Edward Robinson.

COMPETITION AND REPUTATION

Both van de Velde and Robinson/Smith ultimately aimed at founding Protestant Biblical Geography on fieldwork, but did so from different perspectives – the university theologian Robinson from the scholarly angle and the trained military cartographer van de Velde with a distinct evangelical world view. In this, they were competitors who did not shy away from obtaining each other's results for their own purpose and discrediting the other's reputation. Robinson's letters to Eli Smith from 1852, the year of van de Velde's survey and their second trip to the Holy Land, tell of repeated encounters that were not without friction. Apart from meeting twice, it seems they also travelled back to Trieste on the same ship, a fact that van de Velde never mentioned. On the ship, van de Velde had apparently asked Robinson to borrow Eli Smith's notes, purporting that Smith had consented. Robinson later regretted sharing the information and worried that van de Velde was going to profit from it. In the account of their joint excursion to Pella, Robinson also felt that van de Velde had not duly credited them.[235]

Van de Velde in turn claimed that he had found Smith's compass to deviate by three or four degrees in certain areas, which made the resulting maps flawed. He even quoted a letter he had seen at Heinrich Berghaus's in Potsdam that confirmed the instrument's deficiencies.[236]

Yet the other side was not idle neither. In September 1855, the engraver Hermann Eberhardt, whose studio was in Eisenach, wrote to Perthes about a surprise visit:

> Yesterday I had the pleasure of seeing Dr Kiepert at my studio. I barely had time to hide the Palestine drawings from his keen eye and did not manage to shove the proofs into the folder either. He was *very sorry* that I sent you all the drawings yesterday and he raised my hopes of seeing him here again later this year. He mentioned some of his plans to me and if you are interested, I will be happy to share them when I visit Gotha.
>
> Dr Kiepert was so kind to indicate a number of corrections to the map: Chal*s*is must be Chal*c*is and Abyla Lysania*s* must be Lysani*ae*. If you agree, I will implement them immediately.[237]

This leads us back to August Petermann's remark to his employer that van de Velde's map would be a 'renewal' of Robinson's. We can in fact see that van de Velde used as much information as he could for his map and had managed, not least through Heinrich Berghaus, to obtain field notes and other material from his competitors. As this seems to have happened in a somewhat unauthorised manner (but not worse than Kiepert sneaking into the engraver's studio and peeking at the map the former was working on), these sources were not prominently mentioned in the memoir.[238] After all, Robinson and Smith had published their research extensively, so that it sufficed to name their publications as sources. The same letter by Petermann also expressed the publisher's side of the competition: for Petermann and Perthes, publishing the *Map of the Holy Land* with the many valuable sources it included had an advantage over their rival publisher Dietrich Reimer in Berlin, who also published the *Zeitschrift für allgemeine Erdkunde*, and was a competitor to Perthes's *Mittheilungen*.

MAP MAKERS AS BELIEVERS: CARTOGRAPHY AND PROTESTANT ETHIC

It is not always easy to explain why an explorer or researcher chooses a field or region of study. What exactly inspired Charles William Meredith van de Velde, after a career in the Dutch Navy and the colonies, to make a map of Palestine? A map based on modern, scientific principles that had so long been considered a desideratum?

Van de Velde was part of the Dutch *Réveil* movement, a conservative evangelical revival and reaction to the new rationalism in the European Protestant churches. He was a good friend of Ottho Gerhard Heldring, a main protagonist of the *Réveil*, and worked for their foreign mission in the Dutch East Indies after his military career.[239] He was also close to other evangelical free churches, the Evangelical Union and the Geneva Alliance Évangelique as well as the Protestant mission to the Jews. The latter movement was prominent in Protestant circles of the nineteenth century but especially influential in Great Britain. It was part of the multitudes of nineteenth century religious revivals. The underlying concept was that the conversion of the Jews was a prerequisite for the second coming of Christ. In his travel narrative recounting his surveying trip to the Holy Land, van de Velde repeatedly stated this idea: the land was visibly cursed and in decline. Only settling it with Jews converted to Christianity would save it from its sorry state and bring upon it the prophesied return of the Lord. The idea of a mass Jewish settlement in Palestine was initially a Protestant project. This peculiar evangelical Zionism predated the earliest Jewish concepts by almost half a century. In the mid-nineteenth century, European Jews rather strived for full civil emancipation in their homelands. However, the bloody pogroms in Russia in the 1880s brought about a sizeable Jewish Zionist movement.[240]

As Hommo Reenders pointed out, the idea of going to the Holy Land was suggested to van de Velde by Ottho Gerhard Heldring, who had a keen interest in geography and even corresponded with Carl Ritter on van de Velde's Palestine trip.[241] Yet the cartographer's motives surely go deeper. Much points to the core element of Protestantism as an explanation for van de Velde's motives: the obligation for the individual, as formulated in St Paul's Epistle to the Romans, to use one's talents for the best possible benefit of the public.[242] Van de Velde had the skills, the time, the possibility, so it was his religious duty to take on this task. 'Duty' is a word used over and over again both in his travel narratives and the letters to his publisher. In a key passage, he describes visiting Mount Tabor, the traditional place of the transfiguration of Christ:

> My time was limited; I had to either make the top of Tabor a place for meditation and prayer, or to use it as a point of observation for my measurements. The first would have been full of enjoyment for myself; the last promised to be useful for others. The path of duty was thus plain, and I followed it in expectation that the Lord would perform His promise, in making the consolations of the Holy Spirit to abound towards me, independently of site or locality. On a piece of wall at the eastern side of the top, the highest point of the hill, I placed my instrument.[243]

The cartographer was also not above describing the hardships of his duty. Refuse, vermin, 'boils' from long hours in the saddle, inflamed eyes from looking through a telescope in the sunlight and last, but not least, the local peoples: 'A world so full of deceit and intrigue as is this Eastern one was as yet strange and new to me.'[244] When Lorraine Daston and Peter Galison studied the 'techniques of the self' connected to scientific methods in the nineteenth century, they found that self-discipline and hardships were often described in detail as part of the explorer's report, as in van de Velde's travel narrative and the letters to his publisher. A religious undertone, stressing the humility, asceticism and chastity of the heroic scholar showed that mastering scientific techniques was essentially connected to self-control and self-discipline.[245] Also in this respect, van de Velde was both a true Protestant and a typical nineteenth-century explorer.

All in all, the Dutchman managed to accomplish the hitherto impossible by resorting to simple means and making pragmatic decisions: a full map of the region between the Lebanon ranges and the Sinai peninsula based on his own and third party measurements and representing the latest geographical knowledge of his time. Although the map lagged behind in terms of technical possibilities and completeness of its data base, it is a masterful lesson in efficiency, as it achieved the maximum possible result in a very economical way.

The second edition of the *Map of the Holy Land* was generally considered the best of the region until the publication of the large *Map of Western Palestine* by the PEF in 1880.[246] The Nestor of the German-language Palestine research, Titus Tobler, called the map 'decidedly the best that we now know'.[247] The information that had been collated for the map remained valid for over twenty years. Nevertheless it was Heinrich Kiepert, the cartographer of Robinson and Smith, whose Palestine maps became famous until well into the twentieth century. Why did Kiepert become better known than van de Velde if his maps were not superior? Charles William Meredith van de Velde, as we have seen, was not interested in confining himself to just cartography because he felt that this limited him to the cartographic discourse. He probably also felt that his task had been completed with the delivery of his map. While the young Kiepert was only at the beginning of his career,[248] van de Velde abandoned cartography altogether to further devote his time to art and humanitarian causes.

3 August Petermann's Palestine Maps

Bruno Schelhaas

INTRODUCTION

August Petermann never travelled to Palestine personally and we know nothing about any religious motivations within his biography. This fact differentiates him from the two other cartographers in the focus of our project, Heinrich Kiepert and Charles William Meredith van de Velde. At first glance, Petermann is not in the centre of Palestine cartography. Even Titus Tobler does not mention him in his exhaustive bibliography,[1] and Tobler's successor Reinhold Röhricht only lists him three times in his reference book.[2] We can equally find a similar result within Hans Fischer's comprehensive study on the history of Palestine cartography, with only some few references to August Petermann.[3] In spite of all this, August Petermann's published Palestine maps were, as we will see, innovative and influential within the development of Palestine research in the second half of the nineteenth century. As Haim Goren points out, we can divide Petermann's interest in this region into at least four categories:

- Petermann's own Palestine maps;
- his activities in organising exploration and surveys;
- his business in publishing new maps and research results from different scholars; and
- his intensive collecting activities of all kind of geographical information about this region, including many maps.[4]

The focus of this chapter is on the first point mentioned: the Palestine maps produced directly by August Petermann. One question which arises is: what was his interest in mapping this region? Was it commissioned work or personal motivation, or a mixture of both?

The biography of August Petermann was certainly often in focus.[5] The success story of a highly talented cartographer who came from Bleicherode, a small town in the Prussian province of Saxony, far away from any academic centre. This story starts at the beginning of his career when he was promoted by Heinrich Berghaus and ends at the cemetery at Gotha in 1878, as often told. Up to now, August Petermann has been mostly regarded as an extraordinary protagonist of modern geography and cartography, described as an armchair geographer and obsessive organiser of exploration and geographical description of the world. It is easily possible and helpful to connect his life story with geographical places, in particular Bleicherode, Potsdam, Edinburgh, London and finally, and for a long time, Gotha.

In recent years, the state of research has partially changed, as some studies illustrate a new and critical role of Petermann, especially concerning his ventures within Polar research.[6] Nevertheless, there are still some decisive questions regarding his life and work and his importance for the development of our discipline. In particular, we have only little information about his years in Edinburgh and London – and this is exactly the time (around 1847) that Petermann came into contact with maps of Palestine (see Fig. 3.1).

THE POTSDAM YEARS, 1839–45

Petermann's very first printed map was a French map for Alexander von Humboldt's famous *Central Asia*, published 1843 with the title 'Chaînes des montagnes et volcans de L'Asie centrale'. August Petermann was at that time twenty-one years old, and this map publication was just at the beginning the best possible reference in the international community, a first proof of a young cartographer for the world-famous celebrity Alexander von Humboldt. A second early map is 'Die Erzeugnisse des Deutschen Gewerbefleisses in ihrer National-Ausstellung zu Berlin 1844', a very detailed and aesthetic thematic map and masterpiece by the young August Petermann on the occasion of the Great Industrial Exposition of Berlin, 1844. This work was a result of his education at Heinrich Berghaus's Geographical School (*Geographische Kunstschule*) in Potsdam (1839–45), the best institute for cartographic business outside the military in Germany.[7] Together with Heinrich (Henry) Lange and Otto Göcke, Petermann benefitted from an excellent education, with important skills in surveying, cartography, geography, drawing, engraving, lithography and other techniques. Berghaus's experience in the subject and in the market, as well as his contacts with influential people, including Alexander von Humboldt and Carl Ritter, were certainly very helpful in the building phase of the Geographical School. The relationship between Berghaus and Petermann was very close; Petermann's education was free of charge and he was fully integrated in the family and social life of his teacher and mentor. Besides the early map examples, Petermann was most probably engaged in another of Berghaus's projects, the *Physical Atlas*.[8] Finally, Berghaus had an agreement with Justus Perthes publishing

Figure 3.1 August Petermann, engraving
Source: Leibniz Institute for Regional Geography, Archive for Geography, Por-Petermann01

house in Gotha, especially concerning the support of *Stielers Handatlas* (in short simply the *Stieler*). Petermann had possibly already come into direct contact with the *Stieler* in the 1840s. Two decades later, he would become the author of the Palestine map in this famous atlas. Gerhard Engelmann describes the development of Petermann's early career and his connection with Heinrich Berghaus in detail giving us an important hint:

> Kurz vor dem Beginn der Lehrlingsausbildung [of August Petermann] hatte *Berghaus* Auszüge aus den Tagebüchern erhalten, die *Edwin* [sic] *Robinson* auf seiner Reise durch Palästina (1838) geführt hatte. Aus diesen Itinerarien wollte *Berghaus* eine Spezialkarte von Judäa konstruieren. Er ging 'mit großer Liebe' an die Arbeit, konnte sie aber nicht so bald in Angriff nehmen, wie *Robinson* gehofft hatte, und mußte daher die Arbeit während *Petermanns* Lehrlingszeit abbrechen. 'Zwar habe ich den allergrößten Theil der Itinerare fertig konstruirt daliegen, aber diese mühseligen Arbeiten, die mir manche Nacht gekostet haben, sind mir gegenwärtig nur Merkmale der Erinnerung an eine interessante Beschäftigung'. Jedoch veröffentlichte *Berghaus* noch im gleichen Jahre eine Karte von 'Arabia Petraea and Palestine' mit den Reiserouten von *E. Robinson* und *E. Smith.*
>
> (Just before the start [of August Petermann's] apprenticeship, *Berghaus* received the extracts of the diaries of *Edwin* [sic] *Robinson's* travels in Palestine (1838). *Berghaus* wanted to use these itineraries as the source for constructing a special map of Judea. He began the work with 'great love', but not as quickly as *Robinson* had expected and had to cancel the job during *Petermann's* apprenticeship. 'Indeed most of the itineraries have been completed, but this laborious work which took some nights, is actually a memory of an interesting occupation.' However, *Berghaus* still published the map of 'Arabia Petraea and Palestine' with the routes of *E. Robinson* and *E. Smith* in the same year.)[9]

Apart from this fact, there are no indications of any of Petermann's activities concerning Palestine in these years at Potsdam. As we already know, Heinrich Kiepert inherited the job of producing the maps for Robinson some months later.[10]

THE EDINBURGH YEARS, 1845–7

After his apprenticeship and a short time as teacher at Berghaus's school, Petermann moved to the Scottish capital Edinburgh in the summer of 1845.[11] His companion Henry Lange had already moved there some months earlier. The cartographical firm of William & Alexander Keith Johnston[12] was their first employer. Founded in 1825, this publishing house had established itself as one of the leading experts for map making in Scotland and the United Kingdom. Besides John Bartholomew and Son it was the second most important family firm in the Scottish cartography business in this era. Fleet, Wilkes and Withers explain:

> [...] Alexander Johnston, with his brother William [...] founded the cartographic firm of W. & A. K. Johnston in 1825. Along with other family firms such as John Bartholomew and Son Ltd, the company was established at a time when Scotland's cities required different types of maps and plans, and when exploration overseas likewise demanded new maps. New developments in science and in the presentation of scientific understanding from this period also required new types of maps.[13]

The engagement of the two German cartographers was certainly arranged with the help of Heinrich Berghaus. Their main task was to produce the English edition of Berghaus's *Physical Atlas* and Petermann principally worked on the zoological maps.[14] The cooperation between Berghaus and Johnston had already begun some years earlier. It was mainly the German scholar and cartographer Gustaf Kombst[15] who arranged this British-German liaison. The first English maps of Berghaus's *Physical Atlas* were published in 1843 within the *National Atlas of Historical, Commercial, and Political Geography*, followed by a more comprehensive edition in 1848 with the title *The Physical Atlas. A series of Maps and Notes Illustrating the Geographical Distribution of Natural Phenomena* – even in the German edition, there were no Palestine maps. As McCarthy explains, the *Physical Atlas*, Johnston's breakthrough as a scientific map-maker, 'was the first such atlas to be published in Britain'[16] and a number of reprints followed up to the 1870s.

In exactly the same years that Petermann was working for W. & A. K. Johnston, the company was busy preparing three maps of the Holy Land for the two-volume opus *The Lands of the Bible* by the Scottish missionary John Wilson, who travelled in the Holy Land in 1843.[17] The book contains the following maps:

- Map of Arabia Felix, embracing the provinces of Yemen and Hadhramaut, part of the Tehámah, and the peninsula of Aden;
- Map of the peninsula of Mount Sinai and Arabia Petraea, constructed for Dr Wilson's 'Lands of the Bible', by W. & A. K. Johnston, Geographers and Engravers to the Queen, Edinburgh 1847; and
- Map of the Holy Land, with topographical plans of Jerusalem, Tyre, Sidon, Akka, and Beirut. Constructed for Dr Wilson's 'Lands of the Bible', by W. & A. K. Johnston, Geographers and Engravers to the Queen, Edinburgh 1847.

In George Smith's biography of Wilson, we find an interesting source with reference to the Palestine map. Wilson wrote to his missionary companion Robert Nesbit on 6 March 1845:

> I am with Jervis,[18] who is doing great and good things for the East. Yesterday morning he forwarded to Prince Albert, without my knowledge, my proof of the raised map of Palestine. The Prince himself laid it before the Queen, who was much pleased with it,

> and ordered her private secretary to inform us that Her Majesty will graciously accept the dedication of the map from him and Dr. Wilson.[19]

It seems very probable that Petermann was aware of this project, but we have no evidence for his direct involvement in the map production for Wilson's book. We can only find 'W. & A. K. Johnston. Geographers and Engravers to the Queen Edinr.' as author of the maps. Petermann's name cannot be found anywhere within the text or on the map sheets. Wilson refers to the maps in his introduction explicitly, with references to all relevant sources and names of Palestine research and cartography, the British surveys, Joseph von Russegger, Edward Robinson, Ernst Gustav Schultz, to Heinrich Berghaus, Carl Ritter and Heinrich Kiepert. He explains: 'The maps which accompany this work have been constructed by the Messrs. Johnston, geographers to the Queen, under my own personal direction.'[20]

THE LONDON YEARS, 1847–54

In June 1847, August Petermann moved from Edinburgh to London. In September of the same year, he reported to Justus Perthes:

> Seit kurzer Zeit befinde ich mich in der 'Metropolis', wie die stolzen Briten gewöhnlich vorziehen, ihre Hauptstadt zu nennen. Ein Feld verschieden von dem in Edinburgh bietet sich hier den Leuten unseres Faches dar; denn London ist ohne Zweifel der erste Centralpunkt geographischen Wissens wenigstens in Bezug auf die außereuropäischen Theile der Erde, was auch ganz natürlich erscheint, wenn man hinblickt auf die Verfassung des britischen Reiches.
>
> (I have not been very long in the 'metropolis', as the capital is commonly called by the proud Britons. The situation for our subject is quite different to Edinburgh because London is undoubtedly the central point of geographical knowledge, at least for the non-European parts of the globe. This seems to be only natural with regard to the state of the British Empire.)[21]

Petermann's years in Edinburgh and later in London demonstrate a very interesting part of British-German relations in the nineteenth century as described especially by Ulrike Kirchberger, Philipp Felsch and Richard Scully.[22] Scully speaks of 'cartographic freemasonry' and mentions the relation between Johnston and Petermann, but explains in detail the history of relations with the connection between Petermann (and other German cartographers after him) and the Bartholomew publishing house.[23] It was Francis Herbert who first mentioned the quarrel between Johnston and Petermann, which came up some years later about plagiarism and the authorship of some maps.[24] This, however, had no serious consequences for Petermann's later career in Gotha.

It is, however, important to note – and this is nothing new – that Heinrich Berghaus was a key figure and mentor within Petermann's career, much more than Ritter and Humboldt or the Perthes family. Alexander Keith Johnston probably assumed the role as a paternal mentor in Edinburgh and, after that, the Prussian ambassador and scholar Christian Karl Josias von Bunsen[25] served as an important contact point for several issues in London. Within his widespread diplomatic, cultural and scientific engagement, Bunsen was interested in Palestine research and political affairs, too. He played a decisive role for the establishment of the joint Anglican-German Bishopric in Jerusalem in 1841.[26] Altogether, we have to realise a remarkable series of close relations within Petermann's life story, which certainly influenced his subsequent career.

August Petermann's interest in Palestine, the research, the mapping and the source collection, increased immediately after his move to London. His Palestine maps were certainly a product of compilation, combining all available information about the region. This method of operating was common and accepted, and this kind of meticulous work associated him with Kiepert,[27] van de Velde[28] and others of the cartographic community, especially among the Germans. Petermann did not produce any route maps or original maps which visualise actual research results concerning Palestine. This is a crucial difference in comparison with Kiepert and van de Velde.

The very first visualisation of Palestine geography with Petermann's name was the 'Diagram on the Fall of Rivers', published together with his comprehensive contribution and discussion of Edward Robinson's paper in the *Journal of the Royal Geographical Society* (*JRGS*) in 1848 (see Plate 7).[29]

After his move to London, Petermann was immediately busy in networking within the German community in London, but also with the national institutions. The most important was certainly the Royal Geographical Society of London (RGS).[30] Here was the place to encounter the authorities of the geographical community, to study the rich collections of the society and here was the place to obtain new and exclusive information about the exploration of the globe. Altogether, it was an ideal basis to start a cartographic business. The activity of Petermann within the RGS was impressive, especially for a young, foreign scholar. His election as full member had taken place on 22 June 1846, one year before his move from Edinburgh to London. Petermann's main sponsor was certainly his employer and paternal mentor in these years, Alexander Keith Johnston.[31]

On the other hand, Petermann was not the only German member within the RGS. In 1846, we find within the forty Foreign Honorary Members Heinrich Berghaus, Alexander von Humboldt, Carl Ritter and some other prominent scholars; among the twenty-two Corresponding Members were two more Germans. The list of Ordinary Members shows only one identifiable German in this year, the botanist Frederick Scheer.[32]

Petermann took over official duties within the RGS, so at the Anniversary Meeting in 1851 he was one of the 'Scrutineers for the Ballot', and again in 1852 and 1853.[33] Finally, he was appointed as 'Physical Geographer and Engraver on Stone in Ordinary to the Queen' on 18 June 1852.[34] In the first three years in London, Petermann realised map production with several partners, especially with the engraver John Dower and with the publisher William S. Orr.[35] In 1852, he opened his own company, A. Petermann's Geographical Establishment, residing at 9 Charing Cross.[36]

His first maps, which he produced in London from 1848, were thematic maps of the British Isles, but always connected to the actual situation in the United Kingdom, for example the British census maps or the cholera maps, today known as classics of early thematic mapping. Petermann was able to visualise several distributions of phenomena, or almost any kind of statistics in a high quality standard, and within a very short production time.[37]

In connection with the research on Palestine, we find Petermann's name for the second time in the *JRGS* in 1850, with 'Notes on the physical geography of Palestine' by Louis von Wildenbruch,[38] and his own note on the climatological diagram of Palestine.[39]

The public discussion within the RGS about the depression of the Jordan Rift Valley, the exact level of the Dead Sea and the course of the Jordan culminated in 1847–8 with Edward Robinson's paper, read on 22 November 1847.[40] Petermann's reaction followed in the RGS session on 28 February 1848 with his paper 'On the fall of the Jordan, and of principal rivers in the United Kingdom'.[41] Thomas Howard Molyneux's paper about his 1847 expedition to the Jordan and the Dead Sea, read on 27 March 1848, was the closing part of this short but vivid discussion.[42]

In his notebook, Petermann sketched an initial outline of his first article (see Fig. 3.2):

> Entwurf [draft].
>
> On the fall of the Jordan and the fall of English Rivers
>
> I. Reference to Robinson's paper. – His opinion of Symonds measurements. – Russeggers appeal and Wildenbruchs recent barometrical observations.
>
> II. Robinsons parallel drawn betr. Jordan and other Rivers. – On the fall of the Rivers in general. – The size of Rivers, the area, length etc. – Result of my recherches of the hydrography of the British Isles.[43]

In the main part of the 'Diagram on the fall of rivers', Petermann compares the British rivers Dee, Clyde, Tweed and Shannon with the fall of the Jordan, according to the bearings of the British Royal Engineer officer John Frederick Anthony Symonds[44] and of the orientalist and Prussian consul in Beirut, Louis von Wildenbruch.[45] The profile at the bottom 'Section from the Red Sea through El Arabah and El Ghor to Jebel esh-Sheikh' shows the different altitudes according to Wildenbruch, Schubert, Russegger, Robinson, Molyneux and Moresby. Of course, this diagram is not a real map in a strict definition, but it is a very detailed example of visualising newest research data by comparing and interpolating. This method had already been established within nineteenth-century cartography and Petermann was a protagonist of this school. The profile of altitudes was used and improved several times on Petermann's subsequent Palestine maps.

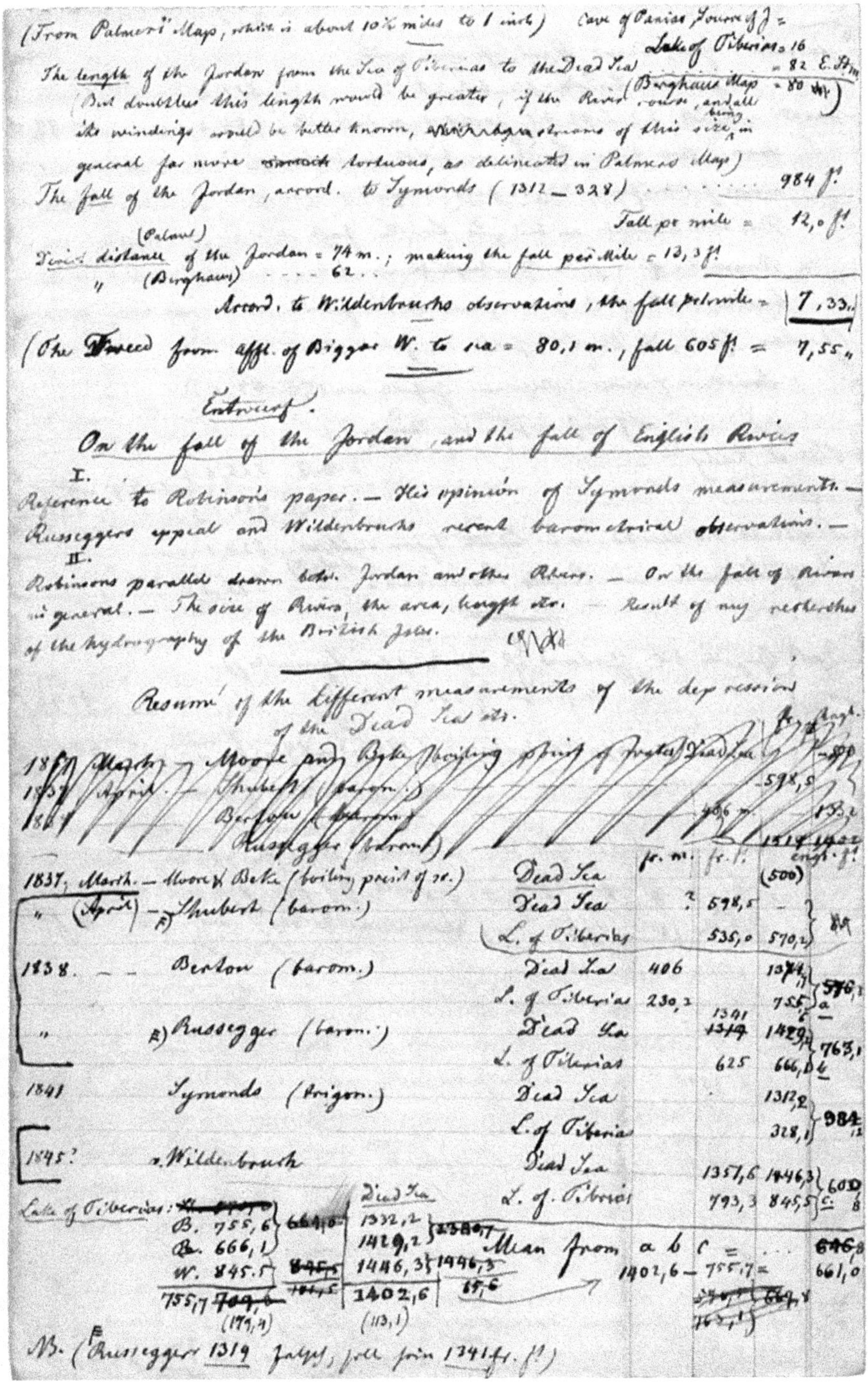

(From Palmer's Map, which is about 10½ miles to 1 inch) Cave of Panias Source of J. = Lake of Tiberias = 16

The length of the Jordan from the Sea of Tiberias to the Dead Sea = 82 E. M.

(But doubtless this length would be greater, if the River course, and all (Berghaus Map = 80 M.) its windings would be better known, ... streams of this size being in general far more tortuous, as delineated in Palmer's Map)

The fall of the Jordan, accord. to Symonds (1312 – 328) .... 984 ft.

Fall per mile = 12,0 ft.

Direct distance (Palmer) of the Jordan = 74 m.; making the fall per Mile = 13,3 ft.

" (Berghaus) ---- 62 ..

Accord. to Wildenbruchs observations, the fall per mile = 7,33 ..

(The Course from affl. of Biggar W. to sea = 80,1 m., fall 605 ft. = 7,55 "

Entwurf.

On the fall of the Jordan, and the fall of English Rivers

I.

Reference to Robinson's paper. — His opinion of Symonds measurements. — Russeggers appeal and Wildenbruchs recent barometrical observations. —

II.

Robinsons parallel drawn betw. Jordan and other Rivers. — On the fall of Rivers in general. — The size of Rivers, the area, length etc. — Result of my researches of the hydrography of the British Isles.

Resumé of the different measurements of the depression of the Dead Sea etc.

~~1837 March — Moore and Beke (boiling point of water) Dead Sea~~
~~1837 April — Shubert (barom.) 598,5~~
~~1838 Bertou (barom.) 406 m~~
~~Russegger (barom.)~~

			Fr. m.	Fr. ft.	Engl. ft.
1837, March	— Moore & Beke (boiling point of w.)	Dead Sea			(500)
" (April)	— Shubert (barom.)	Dead Sea	?	598,5	–
		L. of Tiberias		535,0	570,2
1838	— Bertou (barom.)	Dead Sea	406		1374
		L. of Tiberias	230,2		755
"	Russegger (barom.)	Dead Sea		1341 ~~1319~~	1429
		L. of Tiberias		625	666,1
1841	Symonds (trigon.)	Dead Sea			1312,2
		L. of Tiberias			328,1 } 984
1845?	v. Wildenbruch	Dead Sea		1357,6	1446,3 } 600
		L. of Tiberias		793,3	845,5

Lake of Tiberias: B. 755,6; R. 666,1; W. 845,5 — 755,7 (179,4)

Dead Sea: 1332,2; 1429,2; 1446,3 — 1402,6 (113,1)

Mean from a b c = 1402,6 – 755,7 = ... ~~646,9~~ 661,0

NB. (Russegger 1319 falsch, soll sein 1341 Fr. ft.)

Figure 3.2 August Petermann (*Notes on Palestine*), manuscript (London, 1847)
Source: FBG SPB Konvolut Syrien 1 1847 2

Petermann most probably saw a worthwhile opportunity to promote his own map-making business when participating in the discussion about the Jordan Rift Valley. There was one decisive advantage: Petermann was able to obtain original data – often exclusively, for instance from the British or American surveys (Symonds, Molineux and Lynch), the data from Louis von Wildenbruch, and later from Conrad Schick, as well as some additional material. Today, the unique Perthes Collection[46] holds many of these documents, which Petermann collected in his British years. His ability of arranging exclusive data management was essential for the successful development of his career, but his early Palestine maps were also just a first job for a young freelance cartographer. His knowledge of actual research and exploration subjects, the meticulous collection of all available sources and the strategic knowledge transfer of geographical information were the main characteristics of Petermann's business.[47] The geography of Palestine was one important piece within this establishment, with the African and Polar exploration following very soon afterwards. In spite of all this, Petermann's Palestine maps can be regarded as highly interesting examples of the cartographic visualisation of that region during the initial peak of modern scientific Palestine research and cartography in the mid-nineteenth century.

The source collection was normally the first step of the map-making process. Petermann had already begun this task concerning Palestine geography in 1847 with a detailed extraction and analysis of the selected sources he was able to reach. The comprehensive extract with the heading 'Bemerkungen über die Depression des Todten Meeres' ('Remarks on the Dead Sea Depression'), taken from previously published contributions very impressively documents this.[48] Petermann collected the measurements of the Austrian geologist Joseph von Russegger[49] and of Louis von Wildenbruch;[50] and he created extensive lists with additional levels from Gotthilf Heinrich von Schubert, Jules de Bertou, John Frederick Anthony Symonds, Karl von Raumer and William Francis Lynch.[51] Finally, he calculated his own altitudes. This was the basis of creating a new altitude profile and, as we will see, several new Palestine maps.

Carl Ritter reviewed the contribution in the *JRGS* very favourably in a letter to Petermann from 21 April 1849:

> Mich haben ganz besonders Ihre hydrographischen Entwicklungen interessiert, und ich freue mich ungemein, daß endlich einmal eine gründliche Entwicklung [...] der Flußläufe und Stromsystem durch Sie angebahnt werden wird. Ihre Abhandlung über die englischen Flüsse im Vergleich mit dem Jordan ist so sehr inhaltreich [...]
>
> Es ist sehr zu bedauern, daß mit dem trigonometrischen Nivellement und der Aufnahme durch Symonds etc so sehr für die Veröffentlichung, verzögert wird. Was mag die Ursache hiervon sein. Mit wahrer Ungeduld erwarten wir hier auf dem Continent diese gewiß höchst wichtige Arbeit, und ich insbesondere für die letzte Berichtigung Palästinas und Syriens da ich mit der Geographischen Ausarbeitung dieser Länder gegenwärtig ganz speciell beschäftigt bin.

(I was particularly interested in your hydrographical elaborations and I am very happy that you will finally initiate the thorough survey [...] of the courses of rivers and hydrography. Your contribution on the English rivers in comparison to the Jordan has been very substantial [...]

It is regrettable that the publication of Symonds's trigonometric levelling and survey has been delayed for so long. What might be the reason? We on the continent have been waiting impatiently for this highly important study, especially myself since I have been busy with the geographical description of Palestine and Syria and its final corrections.)[52]

August Petermann responded on 9 June 1849 to Carl Ritter:

Es gelang mir, Symond's [sic] Triangulation im Original-Manuskript zu meiner Einsicht zu bekommen, leider sah ich aber, dass dieselbe nicht auf astronomischen Positionen basiert sei, und zur Berichtigung der Küste nichts enthalte. Ich hielt es deshalb weiter nicht der Mühe werth, um die Erlaubniß einer Durchzeichnung für Sie [Ritter] bei der Admiralität (deren Eigenthum die Aufnahme ist) nachzufragen, sondern schrieb an Admiral Beaufort, einem persönlichen Gönner von mir, mit dem Gesuch, mir etwaige neuen Bestimmungen der Aufnahmen der Syrischen Küste mitzutheilen; in seiner freundlichen Antwort zeigt er mir indeß an, dass von jener Küste noch nichts vermessen sei, und dass ihre nächste Arbeit im Mittelmeer die Aufnahme Ägypten's sein würde. [...] Noch einmal auf die Symond'sche Aufnahme zu kommen, – sollten Euer Hochwohlgeboren eine Copie derselben in anderer Beziehung als zur Küsten-Berichtigung, interessant oder werthvoll sein, so würde ich mit dem größten Vergnügen unverzüglich mich dafür verwenden.

(I had the chance to obtain the original sketches of Symond's [sic] triangulation. Unfortunately, I noticed that it is not based on astronomical positions and contains nothing for the correction of the coastline. Therefore, I decided it was not worth asking the Admiralty (the owner of the sketches) to draw a copy for you. Instead, I wrote to Admiral Beaufort, a personal friend of mine, to inform me about possible new instructions concerning the survey of the Syrian coast. In his favourable answer, he wrote that there are still no surveys of this coast and that the next project in the Mediterranean will be directed at Egypt. [...] Back to Symond's [sic] levelling: if Your Honour is interested in receiving a copy for any other purpose than the correction of the coastline, I would immediately arrange it with the greatest pleasure.)[53]

This source gives an insight into Petermann's practical work during his time in London. We learn about the use of Symonds's original bearings, Petermann's evaluation and his attempts to require more reliable data from the admiralty, here with the help of his personal contact with Francis Beaufort, head of the Hydrographic Office of the Royal Navy. Finally, Petermann supplied a copy of Symonds's bearings to Ritter.

The event is also mentioned in another source some years later. Carl Ritter wrote to Edward Robinson: '[...] sandte ich nun das Blättchen von den Symondschen Surveys, wie ich es durch Petermann habe erhalten können.' ('[...] I sent the sketch of Symonds's surveys which I received through Petermann').[54] The original sketch can be found in the Perthes Collection Gotha.[55]

Petermann's first published Palestine map with his name on the sheet was the 'Map of Lower Egypt, Sinai and Arabia Petraea', issued in 1848 (see Fig. 3.3). It is also mentioned in his *JRGS* article and listed within the accessions to the RGS library.[56] The map is part of the *People's Dictionary of the Bible*,[57] a popular reference book edited by the English Unitarian minister, educational reformer and founder of the Unitarian Home Missionary Board, John Relly Beard[58] in several editions. The dictionary shows strong references to the German theological tradition and refers to other comparable works such as Winer's *Biblisches Realwörterbuch* and Kitto's *Cyclopaedia of Biblical Literature*, both without any maps.[59] In his

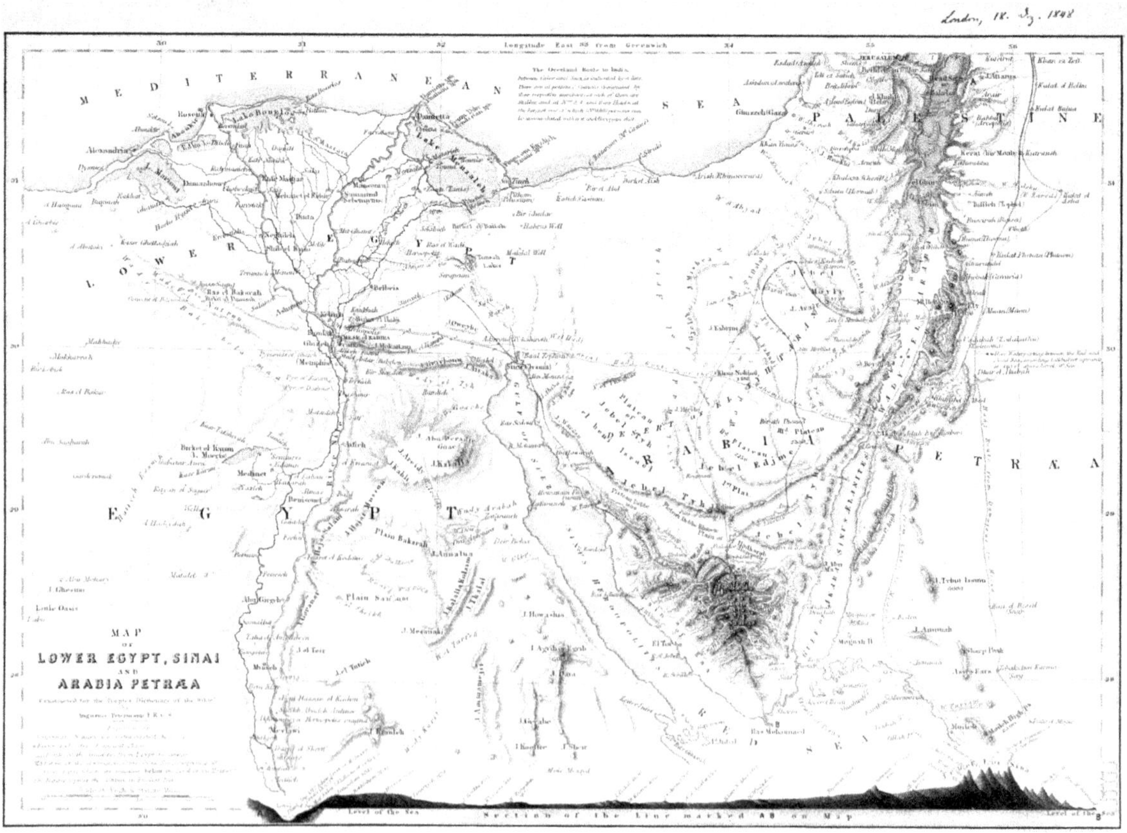

Figure 3.3 August Petermann, *Map of Lower Egypt, Sinai and Arabia Petraea* (London, 1848)
Source: Leibniz Institute for Regional Geography, Central Geographical Library, HK 0444

preface, Beard lists nine main items of the book, including 'II. A Summary of the Geography and Natural History of the Holy Land'. At the end of the introduction, he explains:

> Where an appeal to the eye seemed desirable, wood-engravings, plans, and maps have been supplied; in which, as well as in relation to the materials in general, care has been taken to consult the highest as well as most recent authorities.[60]

The edition history of Beard's dictionary is complicated, but it seems clear that the map production was connected with Johnston in Edinburgh. Volume I of the first edition was published in 1847 by Simpkin, Marshall & Co. in London[61] and volume II followed in 1848.

A second edition cannot be clearly identified in any library, but a third edition came out in 1850 and a reprint already in 1860/61.

The first edition of the *People's Dictionary of the Bible* includes a number of maps constructed and engraved by William and Alexander Keith Johnston in Edinburgh, but we still do not know who the real author or cartographer was. Was it Petermann? Or Alexander Keith Johnston himself? The copy, held in the National Library of Scotland[62] contains three maps in volume I, all without Petermann's name on the sheets:

- The ancient world shewing the origin of nations according to the 1st book of Moses Chap. X, Constructed & Engraved for the People's Dictionary of the Bible by W. & A. K. Johnston, Geographers to the Queen, Edin.r.
- The World as known in the Time of Our Lord, Constructed & Engraved for the People's Dictionary of the Bible by W. & A. K. Johnston, Geographers to the Queen, Edin.r.
- Palestine in the Time of Jesus Christ, Constructed & Engraved for the People's Dictionary of the Bible by W. & A. K. Johnston, Geographers to the Queen, Edin.r.

Volume II contains two sheets:

- No. 1: Plan of Ancient Jerusalem after Kiepert; No. 2: Plan of Modern Jerusalem after Robinson; No. 3: Plan of Modern Jerusalem after Krafft, Constructed & Engraved for the People's Dictionary of the Bible by W. & A. K. Johnston, Geographers to the Queen, Edin.r (see Fig. 3.4).
- Map of Lower Egypt, Sinai and Arabia Petraea. Constructed for the Peoples Dictionary of the Bible by Augustus Petermann F.R.G.S. (see Fig. 3.3).

The second map in volume II is Petermann's 'Map of Lower Egypt, Sinai and Arabia Petraea'. We can discern that Petermann had already joined the first edition of the dictionary and he probably began this work in Edinburgh, but the map is copyrighted not by Johnston, but by Petermann: 'A. Petermann, Cranbourn Str. London'. What does that mean? Petermann was responsible for contributing at least one of his own maps in the dictionary in 1848.

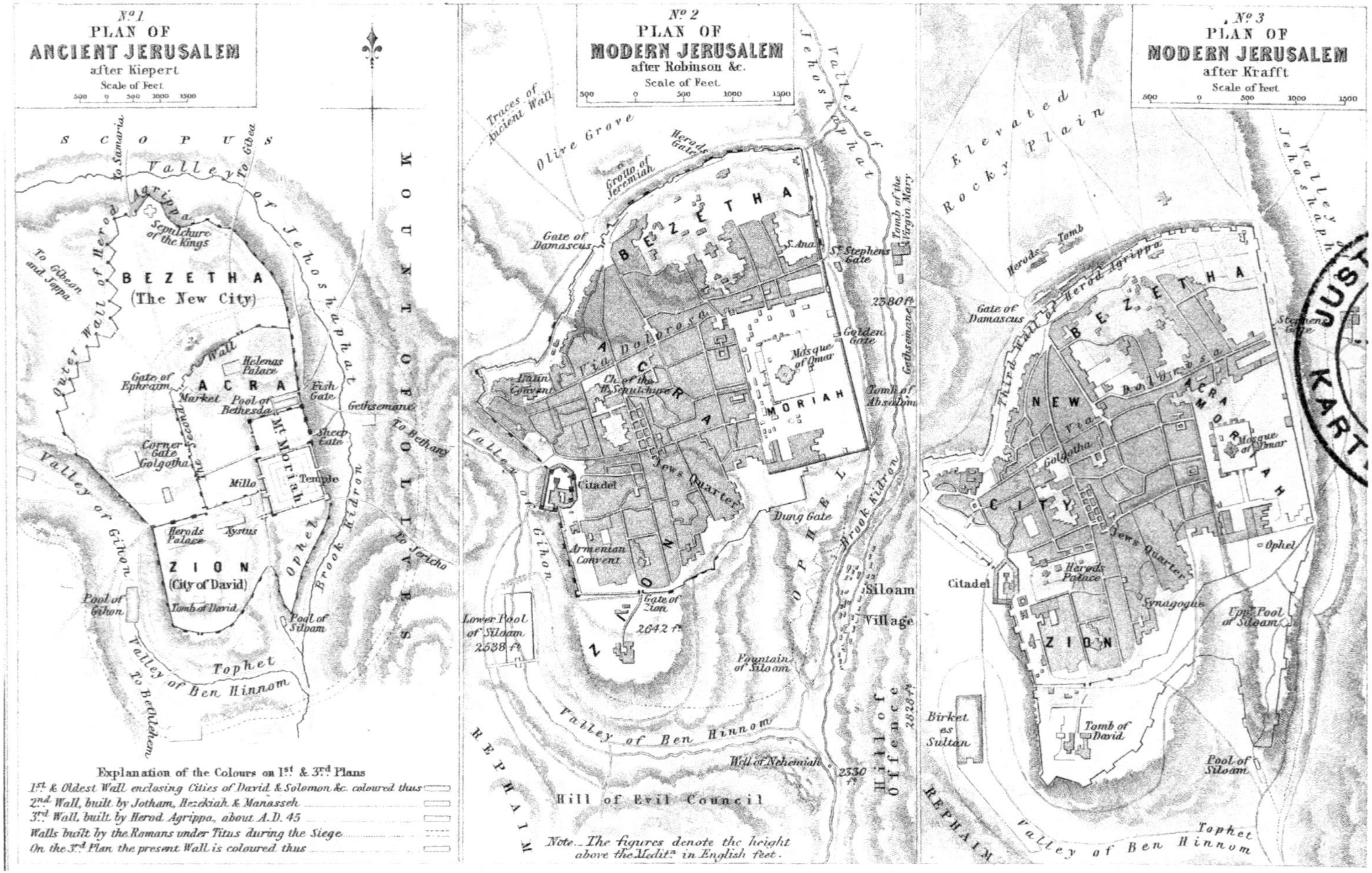

Figure 3.4 Alexander Keith Johnston, *Plan of Jerusalem* (Edinburgh, 1847)
Source: FBG SPK 547-112557201

A first look at the sheet gives us some information about the context of the map. It is a combination of biblical features ('scriptural names', 'ruins and sites of ancient places' or the 'route of the Israelites'), the 'Overland Route to India' (a red coloured line), the 'Hadj Routes' to Mecca, and on the other side with new geographical knowledge ('depression of the Dead Sea'). This synthesis of scriptural and physical geography is typical for the Palestine maps of this era, and Petermann was part of this debate for some years. Although the 'Map of Lower Egypt, Sinai and Arabia Petraea' was part of a religious reference book, Petermann's main interest was on the physical geography of the region. He had already used this early map to illustrate the newest research results and surveys, especially concerning the hypsometric features. One interesting detail on the map sheet underlines this: 'Here Waterparting between the Red and Dead Sea, according to Schubert upwards of 495 ft above Level of Sea'. This refers to the bearings made in 1836 by the multitalented physician, naturalist and philosopher Gotthilf Heinrich von Schubert, a source which was certainly available and of high relevance for Petermann.[63]

At the bottom of the map, Petermann gives a profile of altitudes running from the Dead Sea to Jerusalem through the whole Sinai Peninsula and up to Ras Mohammed at the southern peak. Unfortunately, we do not know how the readers of the dictionary used or interpreted the map.

The Perthes Collection Gotha holds only one short letter from Beard to Petermann, written on 1 September 1848, with information about the position of Kadesh-Barnea[64] according to Carl Ritter and referring to other sources for constructing the map. The letter closes:

> I have not had time to go over the words minutely. I presume You will do so. The spelling should in General be to the best English maps – for instance the Maps issued by the Society for the Diffusion of Useful Knowledge & Robinson; thus Damietta not Damiette. If the ground plan of your Map is right You will have no difficulty in laying down Kadesh Barnea.[65]

In the third edition of Beard's dictionary[66] published in 1850, Petermann already appears as the contributor of corrections and additions of some maps. He probably used Johnston's copper plates or sketches to construct his new maps. We find different sheets of Petermann's maps with these titles in the Perthes Collection, sketch maps with annotations and many signs of usage. Petermann used his old sheets to construct new maps, a quite normal method of cartography. The maps in the third edition are numbered and No. I, III, IV and VI show Petermann's name directly on the map sheet: 3rd edition, London: Simpkin, Marshall & Co. (et al.) 1850; reprint: Manchester: Heywood (et al.) 1861–2:

- No. I: The ancient world shewing the origin of nations according to the 1st book of Moses Chap. X and the Assyrian, Babylonian, Persian and Roman Empires. A. Petermann.

- No. II: The World as known in the time of our Lord.
- No. III: Sketch of the Lake of Galilee, Constructed by Augustus Petermann, F.R.G.S.
- No. IV: Palestine Ancient and Modern, Constructed and Engraved for the Peoples Dictionary of the Bible by W. & A. K. Johnston, Corrections, Additions of Sections &c by Augustus Petermann F.R.G.S. London.
- No. V: No. 1 Plan of Ancient Jerusalem after Kiepert; No. 2 Plan of Modern Jerusalem after Robinson & c.; No. 3 Plan of Modern Jerusalem after Krafft.
- No. VI: Map of Lower Egypt, Sinai and Arabia Petraea. Constructed for the Peoples Dictionary of the Bible by Augustus Petermann F.R.G.S. London 1848.

The 'Sketch of the Lake of Galilee' (see Fig. 3.5)[67] is again a good example of Petermann's cartographic enterprises, which combines and visualises the newest data of scriptural and physical geography. Here, it references the English naval officer Thomas Howard Molyneux's 1847 survey of the Sea of Galilee.[68] In a relatively simple sketch, Petermann gives selected altitudes of some places like 'Safed 2770 ft. (above Meditn.)', 'Nábulus 1672 ft.', 'M^{t}. Gerizim 2470 ft.', 'Jerusalem 2642 ft.' or 'Mt. of Olives 2828 ft.' In the north-west section, he gives more generalised data: 'Heights 300 to 500 ft. above Lake' and in the eastern section 'Mountain Wall 800 to 1000 ft. (above Lake)'. The relief is illustrated with simple hachures around the lake. The height of the sea itself is given as '750 ft. below level of Meditn.', with details of the water depth ('The figures in the Lake denote its depth in fathoms.'). Besides the physical geography, Petermann visualises some biblical features, as we can read in the map legend 'To illustrate Matt. XIV', with 'Probable tracks of the boat' and 'Probable spots where Jesus set sail (1), landed (2) [and] was taken on board'. Finally, two little insets showing the 'Environs of Jerusalem' and the 'Vale of Schechem' allow for comparing the altitudes with the Sea of Galilee in the region of the depression.

The 'Sketch of the Lake of Galilee' was finished by June 1849. Petermann wrote in a letter to Carl Ritter:

> Beiläufig erlaube ich mir ein Exemplar einer Skizze vom Tiberias-See (zu einer theologischen Abhandlung Dr. Beard's gehörig) gehorsamst beizupflichten, in der Hoffnung, daß Ihnen die darin bemerkten, und so viel ich weiß noch nicht von andern Autoren publicirten, von Molyneux' gemachten Reisen, wirklich nicht ganz ohne Interesse sein möchten.
>
> (By the way, please allow me to respectfully present a copy of the Sketch of the Lake of Tiberias (which belongs to a theological work by Dr Beard). I hope that the annotations on Molyneux's journey which, as far as I know, have not yet been published by any author, are useful for you.)[69]

The map No. IV 'Palestine Ancient and Modern'[70] is already listed in the 'Accessions to the Library to 28th May, 1849' in the *JRGS* as donated by August Petermann,[71] some months

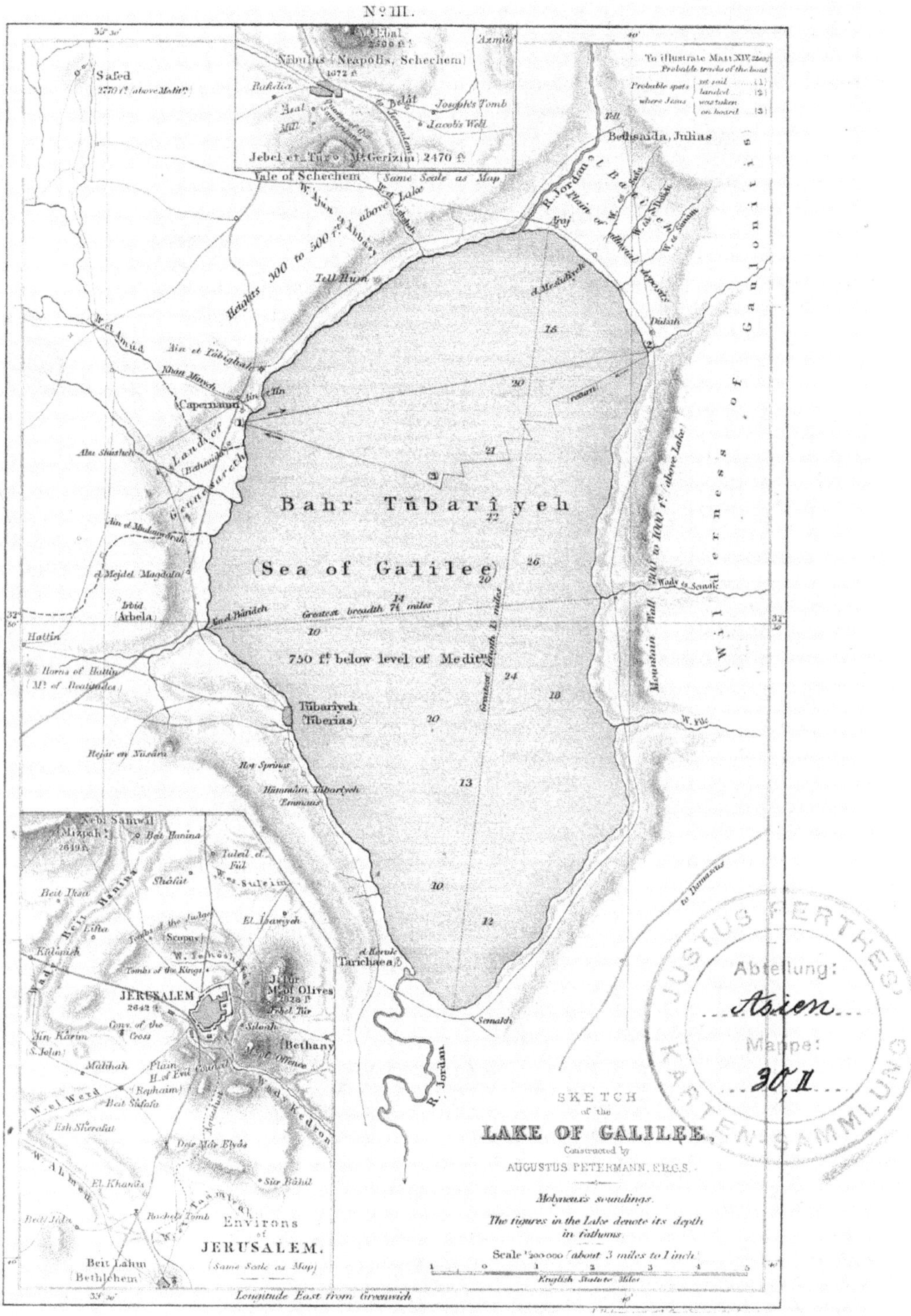

Figure 3.5 August Petermann, *Sketch of the Lake of Galilee* (London, 1850)
Source: FBG SPK 547-112549810

before the issue of the third edition of the dictionary. The donated map is probably part of the unknown second edition, or Petermann produced a single map sheet outside the dictionary for the RGS or others, too. The hand-coloured sheet is again a combination of scriptural and physical geography and shows the ancient boundaries of Palestine (in red), the twelve Tribes, ancient and modern place names and the Jordan Rift Valley in green ('Extent of depression, comprising all those parts which are situated below the level of the Mediterranean'). The map is framed by a very detailed altitude profile with altitudes in English feet. Petermann again gives the level of Tiberias with '750 ft. below Meditn.' and the Dead Sea with '1312 ft. below' as the absolutely deepest point.

Within a short time, John Relly Beard had issued a series of similar publications and some of them again contained the same Palestine maps of the dictionary.[72] We are faced with a number of identical or very similar maps, which were published in several editions and volumes. Beard's *Biblical Atlas* (1849)[73] is a spin-off of the dictionary and contains the same six maps, together with a comprehensive 'introductory sketch of geography' of Palestine and a 'Biblical Gazetteer'. The review in the *Christian Reformer* judges the maps of this atlas very favourably: 'The maps are remarkably good. Although we have seen them before in the author's Dictionary, from the better paper on which they are printed, or from the judicious colouring, they here appear to greater advantage.'[74]

Beard and his publisher, Simpkin, Marshall & Co., wanted to participate in the growing British market of Bible Atlases,[75] especially in competition with William Hughes's *Illuminated Atlas of Scripture Geography*, which had come out in 1840 with a number of valuable maps, and which saw several editions and similar parallel issues. Another series of four Palestine maps by Hughes for the Society for Diffusion of Useful Knowledge followed in 1843.[76] In comparison to August Petermann, Hughes had the advantage of nearly a decade of producing Palestine maps. Consequently, Petermann saw that his only chance in visualising the newest research results of Palestine geography was to focus on the cartography of the extreme physical features of that region.

The most influential Palestine map by August Petermann was his 'Physical Map of Palestine and the Adjacent Countries' (see Fig. 3.6), published in 1850 within the *Atlas of Physical Geography*.[77] It was the first atlas edited and constructed by Petermann, issued in cooperation with the former Congregationalist minister Thomas Milner[78] – and more followed immediately. Petermann's atlas belongs to a series of atlases, constructed by Petermann in a very short time in the early 1850s. The connection with Berghaus and Johnston and their different editions of the *Physical Atlas* is obvious. After his move to London and after having made decisive experiences in this business, Petermann was now able to produce his own atlas. On the cover page of the atlas, he now confidently introduces himself: 'Augustus Petermann, F.R.G.S. Honorary member of the Geographical Society at Berlin; formerly of the Geographical Establishment at Potsdam, and many years assistant in preparing the great Physical Atlas of Berghaus.'[79] The official dedication to Prince Albert is another indicator of Petermann's self-confidence and strategic business marketing.

Jürgen Espenhorst, a leading expert in the field of the history of atlases, explains that he 'is unaware of any comparative study which has yet to be done on the English atlases by Petermann'. He continues: 'It is therefore not possible to provide any further comment here with regard to any changes in map content or cartographic style which may have occurred in publishing these works in Great Britain.'[80]

The reviewer of the newspaper *The Spectator* judged the *Atlas of Physical Geography* favourably:

> And of all the various books we have seen upon the subject, Petermann's and Milner's is the most cleverly popular. The maps are designed in a manner to tell upon the eye as well as to show the phaenomena. The mode of marking the mountain ranges by black lines, the thickness of the line being indicative of the heights, is abandoned for the more picture-like method of shading; which certainly makes a prettier map, and at once says 'This is a mountain.' It strikes us that an effort has also been made to mark the varying phaenomena with more accuracy, by more numerous gradations of shading; which as in the rain-map, may accomplish its purpose, but with the effect of blackness in the deeper parts.[81]

The *Atlas of Physical Geography* is divided into four main chapters (geology, hydrography, meteorology, organic life), accompanied by a series of thirteen thematic world maps. The additional chapters on 'British Islands and Surrounding Seas' and 'Palestine and Adjacent Countries' are connected with the special maps XIV and XV, and the last one is our Palestine map. Milner and Petermann very clearly explain the model of their atlas in the preface:

> Berghaus was the first to propose and execute a representation of the physical features of the Globe in a graphic and pictorial form; and his great work, 'The Physical Atlas', has proved of the greatest utility in the instruction of intellectual nations, besides embodying a vast variety of facts in a compact form, for the use of the scientific classes. Adopting the same plan, but employing original constructions, founded upon the most recent data, and embracing, with two exceptions, the entire surface of the Globe, so as to give a general outline of the system, the present Atlas of Physical Geography is designed to correspond to the existing state of knowledge. The exceptions to the general Maps of the World will be found in the appended Physical Map of the British Isles, which is given on local grounds, and of Palestine and the adjacent countries, which besides possessing some interesting physical phenomena, is given as constituting peculiarly the land of the Bible – the source of our religious faith, and the most ancient fountain of our historical knowledge.[82]

The focus of the 'Physical Map of Palestine and the Adjacent Countries' is on the orography and climatology of the region. Petermann used all the available new data about the altitudes

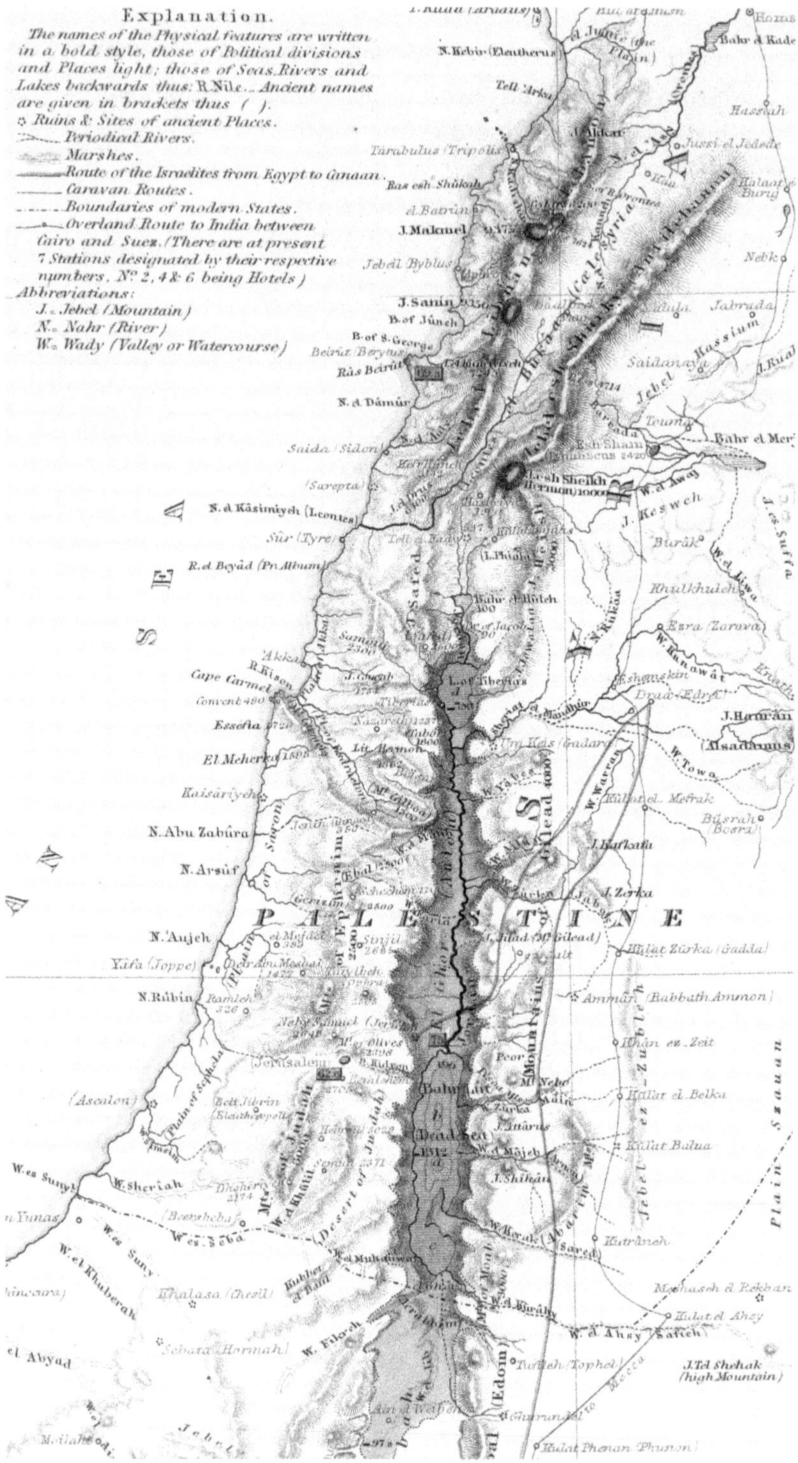

Figure 3.6 August Petermann, *Physical Map of Palestine and the Adjacent Countries* (London, 1850), extract
Source: FBG SPB 2°1010.114

and he coloured the area of depression below the level of the Mediterranean in red; the snow-covered peaks of Lebanon are given in blue. In the explanation on the map sheet, Petermann names some important sources of his map construction. All of them were already known from his earlier Palestine maps: 'The figures denote the height in Engl. ft., being the results of a minute analysis of various observations (many in manuscript) made by: Symonds, Wildenbruch, Russegger, Schubert, Bertou &c.' The quotation once again underlines Petermann's painstaking collection of all available information, including the exclusive and unpublished sources as previously mentioned above.
In the written parts of the *Atlas of Physical Geography*, we can find additional information. The first part of the text is a general geographical description of the region containing astronomical positions, details on the mountains and valleys (with selected altitudes), hydrographical information about seas, the Gulf of Suez and Akaba, as well as rivers and lakes. We also have quotations by Wildenbruch and several biblical references.[83]

The following text is an 'Annotation to the Map', which is a kind of short map explanation.[84] Petermann presents us with an orographic and a climatological regionalisation of this area, the first distinguished into 'region of depression, littoral plains, table-lands, and the Lebanon', the second into 'tropical, warm, temperate, and cold region'. Additional sources of the map production are not given except for the trigonometric surveys of Captain Symonds.

Petermann's 'Physical Map of Palestine and the Adjacent Countries' contains the diagram 'Climatological History of the Months in Palestine &c.' It is a very clear illustration of the climatological year and the vegetation of the region based on a number of original data. The most important information came from Louis von Wildenbruch. Petermann published a description ('Tabular Summary') of the diagram in the *JRGS* (1850). He writes:

> The following *Tabular Summary* I had compiled from published works on Palestine, to accompany a *Climatological* diagram, which has since appeared in the General Physical Map of Palestine in my 'Physical Atlas'.[85]

We can find a few original handwritten drafts of Petermann's diagram in the Perthes Collection. The document with the heading 'Climatological History of the months in Palestine' is nearly equal to the published version in the *JRGS* (see Fig. 3.7).[86] Another sketch is the extract 'Vegetation of Palestine (from Kitto)', which shows several crops of the winter season.[87] The English Biblical scholar John Kitto published several influential and popular books about the Holy Land from the 1830s and he was certainly an important source for Petermann, even in regard to the physical geography.[88]

After Petermann's own climatological calendar, which is equivalent to that on the map sheet, Wildenbruch's detailed corrections and additions follow with the observations from Beirut and the sea coast. Petermann extracted the additions from a long letter, which Wildenbruch wrote to him on 17 April 1849.[89]

Figure 3.7 August Petermann, *Climatological History of the months in Palestine*, manuscript (London, 1850)
Source: FBG SPB 3030 Konvolut Palästina I 2

Parallel to the *Atlas of Physical Geography*, Petermann and Milner published the *Descriptive Atlas of Astronomy, and of Physical und Political Geography*, containing the map 'Palestine Ancient and Modern'[90] and this slightly modified and updated map directly refers to the *People's Dictionary of the Bible* issued some months earlier. Again, we find this map in the list 'Accessions to the Library to 28th May, 1849' in the *JRGS* as donated by August Petermann.[91] Milner and Petermann write in the short explanation of the *Descriptive Atlas*: 'This country [Palestine], a small portion of the Turkish empire in Asia, is made the subject of a special Map, on account of the deep interest attached to it.'[92]

A third atlas, the *Atlas of Political Geography*, was edited by Petermann and Milner in 1851 with the same map 'Palestine Ancient and Modern' and included a similar text explanation.[93] The *Library Atlas of Physical and Political Geography* came out in 1854 at Orr's publishing house in London as a reissue of the *Atlas of Physical Geography* and the *Atlas of Political Geography* in one volume.[94] Finally, we can find the same Palestine map in John Dower's *General Atlas* and *School Atlas*.[95] Besides the atlas maps, Petermann also produced a number of very similar single Palestine maps, some of them for prominent customers such as the National Society.

THE GOTHA YEARS, 1854–78

After Petermann moved to Gotha in the summer of 1854, the series of Palestine maps was not continued in such a dynamic way. Some reprints of his English maps were issued in the late 1850s and 60s, and Petermann was engaged in only some map projects concerning Palestine while working at Justus Perthes in Gotha. One main task for Petermann was now the edition of his new journal, the *Mittheilungen aus Justus Perthes' Geographischer Anstalt über wichtige neue Erforschungen auf dem Gesammtgebiete der Geographie*, in short *Petermanns Mitteilungen*.[96] It was the most important medium of Petermann's successful knowledge communication. The journal immediately became established as a leading platform in cultural and geosciences, not only for the German-speaking world, but also for an international audience. All the renowned names within the geographical community can be found in these volumes. The Journals of the Geographical Societies in Berlin, Paris and London were the only competition. However, Petermann entertained, in part, close connections with all the institutions mentioned, which redounded to his and the company's advantage.

We can find only three Palestine maps in the *Mitteilungen* between 1855 and Petermann's death in 1878 and one profile: Petermann's Map of the Isthmus of Suez (1855), van de Velde's German Map of the Holy Land (1865), the barometric profile by Richard Doergens (1866) and Petermann's map accompanying the railway plan of Conrad Schick (1867). With regard to the neighbouring regions, we can add Petermann's 'Sketch of the Moses' Water Spring in Arabia Petraea and the Suez Bay (1861)' and his 'Map of the Nile Delta and the Isthmus of Suez (1864)'. In contrast to these results, there are many contributions in the journal concerning the region, but without visualisation. The comprehensive research reports in the journal between 1857 and 1865, especially by Johannes Rudolf Roth, Titus Tobler and Charles William Meredith van de Velde impressively document this.[97]

The very first volume of the journal (1855) contained a Palestine map directly connected to Petermann, the 'Kartenskizze des Isthmus von Sues' ('Sketch Map of the Isthmus of Suez', see Fig. 3.8).[98] This map illustrates Petermann's article about the planning of the Suez Canal in the same volume of the journal.[99] We can find a clear combination of two issues here: the actual planning of the Suez Canal and once again the altitudes of the region, especially the depression of the Jordan Rift Valley. Both the text and the map are consequently divided into two parts: the very accurately drawn Nile Delta in the west and Palestine in the north-east. In between, we can see desert with relatively low information density showing some *wadis* and mountain ranges. Obviously, Petermann took the chance to use and update his rich material on the physical geography of Palestine on the occasion of the debate on the Suez Canal and the ventures of Ferdinand de Lesseps and his company.[100] He opens the first chapter on the Isthmus of Suez with an explanation of the most important geographical question, the challenging realisation of the canal:

> Unzweifelhaft die wichtigste geographische Frage, die in dem eben scheidenden Jahre (1855) näher zur Erörterung gekommen ist, als zuvor, und die lebhafte Aufmerksamkeit und Theilnahme Vieler beansprucht hat, ist die Durchstechung des Isthmus von Sues behufs Anlage eines Schiffs-Kanals.
>
> (Undoubtedly, the most important geographical question in the passing year (1855), and which came closer to the argument than ever before as well demanding the lively attention and participation of many, was the breakthrough of the Isthmus of Suez for the purpose of a ship canal system.)[101]

Petermann very critically describes in detail the status quo of Lesseps's plans. He comments on the official measurements and publications of the Suez Company and discusses the different proposed routes of the canal. In particular, he criticises the lack of existing maps:

> Wir sind nicht im Stande gewesen, die Positionen der wenigen Punkte dieses Nivellements auf den uns zu Gebote stehenden Karten zu ersehen [...] Dies ist ein Mangel, den wir uns berechtigt fühlen zu rügen. Aber noch viel mehr müssen wir es tadeln, dass sich auf der dem Lesseps'schen Bericht beigegebenen Karte ganz andere Höhen verzeichnet finden [...]
>
> (It was not possible for us to identify the positions of the few points within this levelling on the maps at our disposal. [...] We feel authorised to criticise this deficiency. To a greater degree, we have to reproach the completely different altitudes found on the map which accompanies Lesseps' report [...])[102]

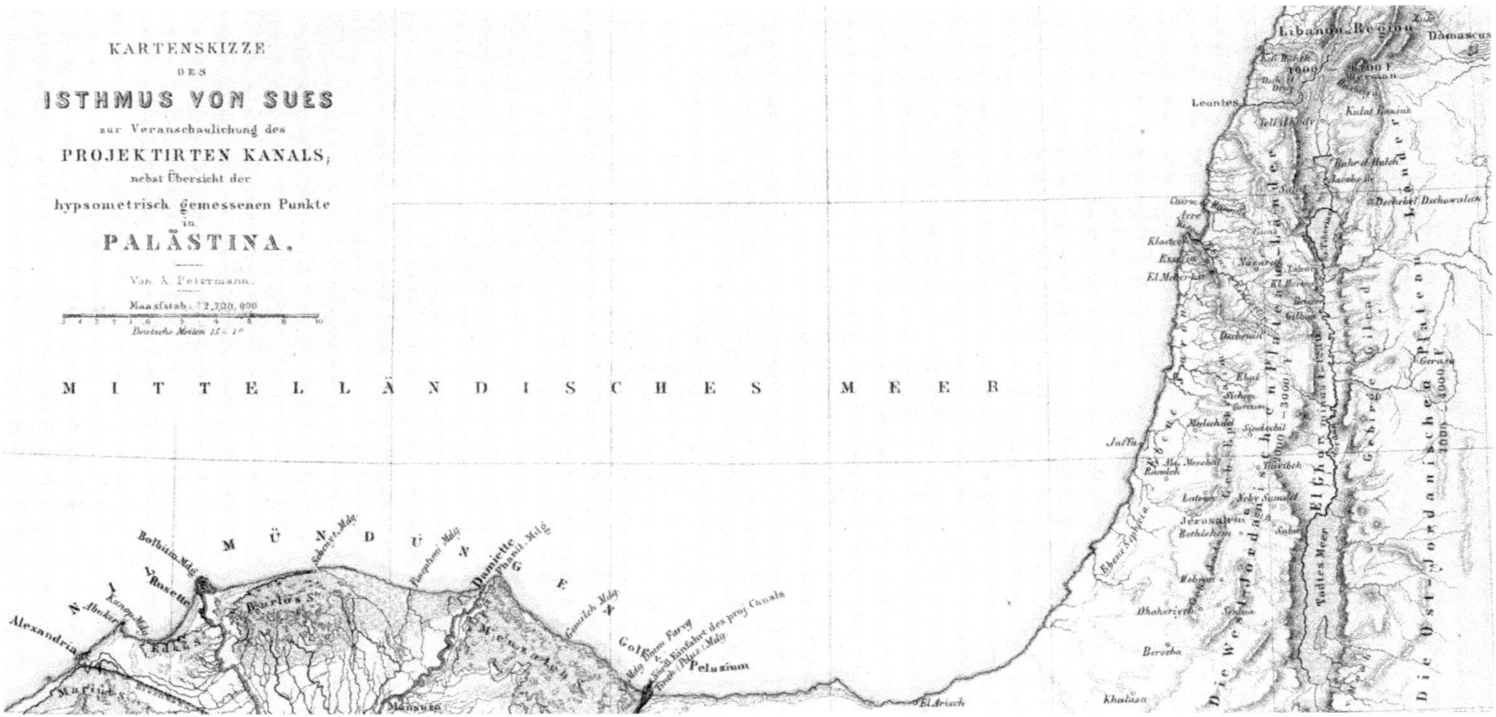

Figure 3.8 August Petermann, *Kartenskizze des Isthmus von Sues* (*Sketch Map of the Isthmus of Suez*) (Gotha, 1855)
Source: FBG SPA 4° 100, Bd. 1, Tafel 23

This critique was certainly Petermann's main argument to produce a newly updated and highly scientific map on this subject, a strategy which would become the successful feature of the *Mitteilungen* for the following years.[103]

The map shows the railway from Alexandria to Cairo and the postal road from Cairo to Suez in red, and the planned canal from Tineh at the Mediterranean coast to Suez at the Red Sea in blue. In the textual part of the second chapter on the hypsometric research in Palestine, Petermann discusses several measurements and presents a detailed list of altitudes according to the previously known results of Bertou, Symonds, Schubert, Russegger and Wildenbruch, and the newest measurements of the British Captain William Allen[104] and the Irish missionary Josias Leslie Porter.[105] The arrangement of the data shows a regional division into Lebanon, Coele-Syria and Leontes Valley, Ante-Lebanon to Damascus, the depression area and the West Jordan Land. Petermann closes his contribution not without modesty and with an appeal for advanced research:

> Die vorstehende Tabelle enthält ziemlich alle wesentlichen Beobachtungen, die man bisher in Palästina angestellt hat, und sie bildet, unseres Wissens, den ersten Versuch, dieselben kritisch zusammenzustellen. [...] Wenigstens darf man hoffen, dass diese provisorische Zusammenstellung das Gute haben wird, den kläglichen Zustand unserer Topographisch-Physikalischen Kenntniss des Heiligen Landes frappant vor Augen zu legen, und so vielleicht Etwas beizutragen, um die zukünftigen Forscher desselben zu veranlassen, dass sie zur besseren Kenntniss das Ihrige thun werden.
>
> (The preceding table contains pretty much all the essential observations concerning Palestine and it is as far as we know the very first attempt to summarise them critically. [...] At least, one would hope, this temporary compilation will have the positive effect of illustrating the poor topographical and physical knowledge of the Holy Land. Maybe it will motivate future researchers to play their part in improving this knowledge.)[106]

Petermann's second Palestine map in the *Mitteilungen* is the 'Originalkarte des Theiles von Palästina zwischen Jaffa und Jerusalem' ('Original Map of Palestine between Jaffa and Jerusalem'),[107] published in 1867, together with Conrad Schick's study about the railway planning between Jaffa and Jerusalem.[108] The project of a railway connection between Jaffa and Jerusalem was vividly discussed in the 1860s and it was especially forced by the Prussian railway engineer, visionary and esoteric Charles Franz Zimpel. In his journal article, Schick explains Zimpel's idea in detail. The railway was finally realised in 1892.[109]

The German multitalented missionary Schick had moved to Jerusalem in 1846, where he worked for Saint Chrischona Missionary, the pietistic project initiated by the Suisse Christian Friedrich Spittler. Since 1850, he carried on his job for the London Society for Promoting Christianity Amongst the Jews (London Jews Society) and he became established as one of the leading experts and a contact person for all scientific, missionary and cultural affairs in Jerusalem and the Holy Land.[110]

The printed map is based on a hand-drawn and coloured sketch by Conrad Schick, dated 1864.[111] Schick's sketched map is meticulously worked out with many settlements, place names and bodies of water, as well as different routes for the proposed roads and railways and with an impressive cartography of the relief with hachures. He writes in the remarks on the map sheet:

> Da der Zweck dieser Karte ist, die von Jaffa (dem Meer) nach Jerusalem führenden Straßen, und diesem Landstrich genau in einzelne, in Hinsicht auf etwa eine neu anzulegende Straße oder Eisenbahn anzugeben, so war nöthig dies im großen Maasstab auszuführen. Da aber Jaffa viel nördlicher als Jerusalem liegt, so hätte das Papier sehr breit sein müssen, um die Längen u Breitengrade mit demselben recht wikkelig [sic!] von einer Seite zur andern zu ziehen, und habe daher vorgezogen, dieselbe richtig aber schief über die Karte laufend einzuzeichnen; ist aber jedem bei genauer Einsicht leicht verständlich. Bei Reduction der Karte kann es dann auf die übliche Weise gemacht werden.
>
> (The purpose of this map is to show the roads from Jaffa (by the sea) to Jerusalem and exactly this stretch of land in particular with regard to a newly proposed road or railway. Therefore, it was necessary to work this out in large scale. Since Jaffa is located far more to the north than Jerusalem, the sheet would have been much broader in order to construct the longitudes and latitudes with the same right angles from one side to another. We decided instead to draw them correctly, but crooked on the page; it is easily understandable for everybody looking at the map carefully. It can be done in the usual way when reducing the map.)[112]

Petermann used the draft sheet in order to work out the map for the journal. The accordance between Schick's sketch and the final printed map is very high. This underlines the high quality of the work by the amateur cartographer Conrad Schick. After studying the first map sketch, Schick sent his comprehensive comments with very detailed corrections back to Justus Perthes with specifications of positions, the settlement, the different routes, the size of several elements, place names, etc. He wrote at the beginning of his remarks:

> und will uns hier in aller Bescheidenheit aufzählen was ich glaube, daß nicht ganz richtig ist, oder nachgetragen oder verbessert werden möchte. Wobei es hier ganz natürlich den Umständen u. Herrn Dr. Petermann überlassen bleibt, wie weit davon noch Gebrauch zu machen ist.
>
> [...](and in all modesty, I wish to present the points which I believe are not fully correct or which have to be added or improved. The realisation should, of course, be left to the circumstances of how much there still is to be done and on Dr Petermann.)[113]

In the journal article, Schick explains the main sources of his map and gives some important details about the map making process:

> Die Lage der Hauptorte habe ich meist nach Van de Velde angenommen und dessen Karte im Allgemeinen dieser vorliegenden zu Grunde gelegt, dann aber den Erfund meiner eigenen Aufnahmen, so wie die Anderer, wo ich wusste, dass sie genauer als die Van de Velde's sind, eingezeichnet. Die Richtungen der Thäler, Wegen und die Lage der Ortschaften habe ich stets durch Winkel mit einem guten Kompass und die Distanzen nach der Uhr gemessen. Die Karte stellt das Land zwischen Jaffa und Jerusalem in einer solchen Breite dar, wie dieselbe nur bei Anlegung einer Strasse in Betracht kommen kann. [...]
>
> Mehrere meiner Wanderungen, wie sie auf der Karte angegeben sind, bedürfen keiner weiteren Erklärung.
>
> (I mostly oriented the positions of the main places according to Van de Velde. His map was the basis in general and I drew on my own measurements as well as others where I knew they were more reliable than Van den Velde's. The directions of the valleys, roads and the positions of the settlements are constructed with the help of an angle with a good compass and the distances measured with a clock. The map shows the country between Jaffa and Jerusalem in such a broad way due to road planning. [...]
>
> Several of my travels, as given in the map, need no further explanation.)[114]

Besides the *Mitteilungen*, and many years before, it was *Stielers Hand-Atlas* which had served as a very successful main product of Justus Perthes.[115] After the Napoleonic Wars, the cooperation between Adolf Stieler and the publisher Justus Perthes developed fruitfully; the idea of a new and high quality atlas was born and the first map sheet was issued in 1816. The single map sheets and whole editions were constantly updated, but the complete first edition with 75 sheets was not finished until 1831. Altogether the *Stieler* saw eleven editions between 1817 and 1945, and a number of foreign editions, school editions, pocket editions and other spin-offs, with hundreds of sheets and a great number of cartographers, drawers, engravers and many more of Justus Perthes's collaborators who participated in this core project and top seller of the publishing house.

In the early editions, Palestine was not the focus of Adolf Stieler and his team. The sheet 42/43 'Das Mittellaendisch und Schwarze Meer' ('Mediterranean and Black Sea'), worked out by Christian Gottlieb Reichard and initially published in 1818 and revised in 1833, showed only a small and rough Palestine inset. The first single sheet for Palestine came out in 1844 as sheet no 42b 'Palaestina nach den zuverlässigsten alten und neue Quellen' ('Palestine according to most reliable ancient and modern sources'). This map was elaborated on by Karl von Raumer after a long and meticulous preparation and source collection and saw several revised reprints up to 1868.[116] In the following editions after the first issue of 1844, we find Friedrich von Stülpnagel as the co-author and Hermann Berghaus, the nephew of Heinrich Berghaus, as the drawer of the sheet. Stülpnagel was the most important collaborator of the atlas after the era of Adolf Stieler, who died in 1836. After Stülpnagel's death, August Petermann inherited this position as the main editor responsible for the *Stieler* in 1865.

The Palestine map is surrounded by six insets: 'Die Länder der Heiligen Schrift' ('The countries of the Holy Scripture'), 'Umgebungen von Jerusalem' ('Environs of Jerusalem'), 'Jerusalem zur Zeit der Zerstörung durch Titus' ('Jerusalem in the time of destruction by Titus'), 'Jerusalem zur jetzigen Zeit' ('Modern Jerusalem'), 'Zug der Israeliten aus Aegypten nach Kanaan' ('Exodus of the Israelites from Egypt to Canaan') and 'Höhendurchschnitt vom Sinai bis zum Gr. Hermon' ('Altitude Profile from Sinai to Great Mount Hermon').

In the short textual part within the atlas ('Bericht über die Blätter des Hand-Atlas') ('Report on the pages of the Hand-Atlas') we find some details about the sources Karl von Raumer used for his map:

> Unter den Karten von Palästina benutzten wir vorzüglich die von *Berghaus* und *Kiepert*. Was *Berghaus* für das ostjordanische, das leistet *Kiepert* für das westjordanische Land, indem Beide mit großem Geschick und gewissenhafter ausdauernder Mühsamkeit den ausgezeichneten Reiseberichten folgten, *Berghaus* denen von *Burckhardt*, *Kiepert* denen von *Robinson* und *Smith*.
>
> Unsere Karte ist vorzugsweise bestimmt, zum Verständnis der Bibel zu dienen.
>
> (Among the maps of Palestine, we primarily used ones by *Berghaus* and *Kiepert*. What *Berghaus* did for the lands east of the Jordan River, Kiepert did for the lands west of the Jordan, in which both followed the distinguished travel accounts with great skill and care and persevering labour. *Berghaus* used, in particular, *Burckhardt* for mapping Transjordan and *Kiepert* used *Robinson* and *Smith* for the lands west of the Jordan.
>
> The intention of our map is mainly directed at understanding the Bible.)[117]

Raumer's most important source was, of course, his own work on Palestine, published in 1835. It is the voluminous historical-geographical study of a scholar who never travelled to Palestine personally.[118] The third edition of the Palestine map, issued in 1859, shows Petermann's name for the first time: 'mit Zusätzen von A. Petermann 1855' ('with additions by A. Petermann 1855'). Petermann added especially important features of the physical geography. In the explanation on the map sheet, he introduces four geographical areas of Palestine: the littoral area, the depression area, the plateaux area and the Lebanon mountain range. He explains: 'Die Höhenangaben beruhen auf sorgfältigen Untersuchungen der bisherigen Messungen [...]' ('The elevations are based on careful studies of the existing measurements').

In the fifth edition (1868), we finally find Petermann's new map 42b as a double sheet 'Palästina' ('Palestine'), together with 'Das Libanon-Gebiet' ('Lebanon') and the inset 'Jerusalem und Umgegend' ('Jerusalem and its environs'), engraved by Eberhardt and Grunert.[119] In Petermann's explanation in the *Mitteilungen* we find some information about the new map and its creation:

> Nicht weniger als 350 Höhenzahlen sind zur Unterstützung und zum Beleg unserer Höhenkunde auf der Karte eingetragen, und zwar sind diese Höhenzahlen das Endresultat einer sehr sorgfältigen und eingehenden Diskussion der zahlreichen bisherigen, oft wenig übereinstimmenden, Messungen verschiedenen Werthes, von der Englischen Nivellirung von Jaffa über Jerusalem zum Todten Meer, die in diesem Gebiete 55 Höhen auf den tausendsten Theil eines Fusses genau angiebt, bis auf barometrische und thermometrische Messungen, die von der Wahrheit mitunter um mehrere 100 Fuss abweichen. [...] einstweilen sei erwähnt, dass für das Hauptgebiet Palästina's, das West-Jordan-Land, die Nivellirungen der Engländer, für das Libanon-Gebiet deren trigonometrischen Aufnahmen, für das Ost-Jordan-Land die sorgfältigen barometrischen Messungen vom Ingenieur R. Doergens die Grundlagen zur Korrektion aller übrigen Höhenbestimmungen abgaben.
>
> (No less than 350 height measurements are given in the map to illustrate the status quo of the orography. These altitudes are the final result of a very detailed and thorough discussion of the numerously available, but often contradictory measurements of varying quality by the English levelling from Jaffa via Jerusalem to the Dead Sea, which shows 55 altitudes with an accuracy of a thousandth part of a foot, save for barometric and thermometric measurements which occasionally deviate from the truth by several hundred feet. [...] Meanwhile it should be mentioned that the engineer R. Doergens provided the basis for correcting all the rest of the height definitions for the main area of Palestine, the West Jordan Land, the English levelling, for the area of Lebanon and its trigonometric recordings and for the East Jordan Land.)[120]

A SHORT SUMMARY

August Petermann had been participating in the debate on the geography of Palestine since the late 1840s. His focus was always on the physical geography, the altitudes, hydrography, climatology and vegetation of the region. Even in the maps belonging to religious books and atlases, the features of physical geography are obvious. Petermann was able to benefit from his excellent contacts in London, among which were the Royal Geographical Society and the Royal Navy. Within these networks, he participated in the short but very vivid discussion on the depression of the Jordan Rift Valley. We can classify Petermann's Palestine maps into British- or Gotha-based or into maps for religious publications, atlases, journals and single maps. The early maps were part of his commissioned work as a freelance cartographer and part of his own atlas projects. His later cartographic work within the subject was part of his position at Justus Perthes publishing house, where he acted as journal and atlas editor, as cartographer responsible for specific maps and, most of all, as organiser and merchandiser of geographical knowledge in the widest sense. The geography of Palestine was a small part within Petermann's oeuvre, but he can be regarded as an influential mediator between several involved actors of Palestine research between 1850 and 1878. His close connections with Heinrich Berghaus, the Royal Navy (in particular John Washington, Francis Beaufort, Thomas Howard Molineux and John Frederick Anthony Symonds), German-speaking scholars Louis von Wildenbruch, Joseph Russegger, Titus Tobler, Johannes Rudolf Roth, Gotthilf Heinrich von Schubert and Conrad Schick, and with Charles William Meredith van de Velde and other relevant participants of the Palestine research community are impressive and unique. Petermann's Palestine maps are the visualised synthesis, based on all available and often exclusive sources and elaborated in high quality cartography.

Table 3.1 Petermann's Palestine Maps

Title	*Reference*
Diagram on the Fall of Rivers by Augustus Petermann, with section from the Red Sea through El Arabah and El Ghor to Jebel esh-Sheikh (see Plate 7)	*JRGS* 18 (1848)
Map of Lower Egypt, Sinai and Arabia Petraea, constructed for the Peoples Dictionary of the Bible by Augustus Petermann F.R.G.S. [ca. 1:2,800,000] (see Fig. 3.3)	Beard, *The People's Dictionary of the Bible*, 1847/8 (1st edn), 2 vols; 1850 (3rd edn) Beard, *A Biblical Atlas*, 1849
No. I: The ancient world shewing the origin of nations according to the 1st book of Moses Chap. X and the Assyrian, Babylonian, Persian and Roman Empires by Augustus Petermann	Beard, *The People's Dictionary of the Bible*, 1850 (3rd edn) Beard, *A Biblical Atlas*, 1849
No. III: Sketch of the Lake of Galilee, constructed by Augustus Petermann, F.R.G.S. (see Fig. 3.5)	Beard, *The People's Dictionary of the Bible*, 1850 (3rd edn) Beard, *A Biblical Atlas*, 1849 Beard, *Scripture Vindicated Against Some Perversions of Rationalism*, 1849
No. IV: Palestine Ancient and Modern, constructed and engraved for the Peoples Dictionary of the Bible by W. & A. K. Johnston, Corrections, Additions of Sections &c by Augustus Petermann F.R.G.S. London	Beard, *The People's Dictionary of the Bible*, 1850 (3rd edn) Beard, *A Biblical Atlas*, 1849
No. XV: Physical Map of Palestine and the Adjacent Countries, constructed by Augustus Petermann, F.R.G.S, engraved by I. Dower, Pentonville, London [1:3,900,000] (see Fig. 3.6 and Plate 8)	Milner and Petermann, *The Atlas of Physical Geography*, 1851 Milner and Petermann, *The Library Atlas of Physical and Political Geography*, 1854
No. XXX: Palestine Ancient and Modern, constructed by Augustus Petermann, F.R.G.S, engraved by John Dower, Pentonville, London	Milner and Petermann, *A Descriptive Atlas of Astronomy, and of Physical and Political Geography*, 1850 Milner and Petermann, *The Atlas of Political Geography*, 1851 Milner and Petermann, *The Library Atlas of Physical and Political Geography*, 1854 Dower, *A New General Atlas of Modern Geography*, 1860 Dower, *School Atlas of Modern Geography*, 1852

Title	*Reference*
Physical Map of Palestine and the Adjacent Countries, constructed by Augustus Petermann, F.R.G.S.	Religious Tract Society, *The New Biblical Atlas and Scriptive Gazetteer*, 1852
Palestine and the Adjacent Countries, engraved on stone by Augustus Petermann. Published by the National Society. [London]: National Society, 1852	
Palestine illustrating its Geographical, Physical and Historical Features, drawn by Augustus Petermann F.R.G.S. with the co-operation of the Revd. S. Clark. M.A. [London: National Society 1853]	
Bl. 42b: Palaestina. Nach den zuverlässigsten alten und neuen Quellen. Von K. v. Raumer und F. von Stülpnagel. Mit Zusätzen von A. Petermann. [ca. 1:1,060,000]	Stieler, *Hand-Atlas über alle Thiele der Erde*, 1855, 1856, 1858
Kartenskizze des Isthmus von Sues zur Veranschaulichung des projektirten Kanals. Nebst Übersicht der hypsometrisch gemessenen Punkte in Palästina. Von A. Petermann [1:2,700,000] (see Fig. 3.8)	PGM 1 (1855)
Originalskizze der Moses-Quellen im Peträischen Arabien. Aufgenommen von Th. v. Heuglin, 27. Mai 1861. E. Debes del.; Tafel 15: Karte der Bai von Súes. Nach den Engl. Aufnahmen v. Comm. Mansell, 1856. Von A. Petermann	PGM 7 (1861)
Bl. 42b: Palästina im Massstabe von 1:200,000 Von A. Petermann / Das Libanon-Gebiet im Maasstabe 1:500,000 / Jerusalem und Umgegend im Massstabe von 1:150,000 Gest. v. Eberhardt & Grunert (see Plate 10)	Stieler, *Hand-Atlas über alle Thiele der Erde*, 1868
Originalkarte des Theiles von Palästina zwischen Jaffa und Jerusalem. Mit besonderer Berücksichtigung der vorhandenen und projektirten Communicationen, nach Originalzeichnungen und mit Benutzung der neuesten Aufnahmen und Forschungen von Conrad Schick. A. Petermann dir. (see Plate 9)	PGM 13 (1867)
Bl. 61: Palästina im Massstabe von 1:200,000 Von A. Petermann / Das Libanon-Gebiet im Maasstabe 1:500,000 / Jerusalem und Umgegend im Massstabe von 1:150,000 Gest. v. Eberhardt & Grunert	Stieler, *Hand-Atlas über alle Thiele der Erde*, 1876

Notes

CHAPTER 1

1 Titus Tobler, *Bibliographia geographica Palaestina. Zunächst kritische Übersicht gedruckter und ungedruckter Beschreibungen der Reisen ins heilige Land* (Leipzig, 1867), pp. 162–3 ('the accepted reformer in the geography of Palestine'); Frederick Jones Bliss, *The Development of Palestine Exploration* (London and New York, 1906), pp. 184–223 (p. 205: 'Friend and foe, however, recognized in it an epoch-making work'); Yehoshua Ben-Arieh, *The Rediscovery of the Holy Land in the 19th Century* (Jerusalem, 1979, 2nd edn 1983), pp. 85–91 (85: 'The foundation of Palestine Exploration'); Neil Asher Silberman, *Digging for God and Country. Exploration, Archaeology, and the Secret Struggle for the Holy Land, 1799–1917* (New York, 1982), pp. 37–47 ('Scientists and geographers, as well as Biblical scholars and clerics, hailed his achievement').

2 Henry Boynton Smith and Roswell Dwight Hitchcock, *The Life, Writings and Character of Edward Robinson* (New York, 1863, repr. New York, 1977); Bliss, *Development of Palestine Exploration*, pp. 184–223; Ben-Arieh, *Rediscovery*, pp. 85–91; Haim Goren, 'Carl Ritter's contribution to Holy Land research', in A. Buttimer, S. D. Brunn and U. Wardenga (eds), *Text and Image. Social Construction of Regional Knowledge* (Leipzig, 1999); Jay Gomer Williams, *The Times and Life of Edward Robinson. Connecticut Yankee in King Solomon's Court* (Atlanta, 1999); Renaud Soler, *Edward Robinson (1794–1863) et l'émergence de l'archéologie biblique* (Paris, 2014).

3 Smith and Hitchcock, *Life, Writings and Character of Robinson*, pp. 40–8; Ussama Makdisi, *Artillery of Heaven. American Missionaries and the Failed Conversion of the Middle East* (Ithaca and London, 2008), pp. 58–61; Jerry Wayne Brown, *The Rise of Biblical Criticism in America, 1800–1870. The New England Scholars* (Middletown, CT, 1969), pp. 45–9; Jay Gomer Williams, *The Education of Edward Robinson* (New York, 1997), pp. 4–6; Jay Gomer Williams,*Times and Life*, pp. 81–94.

4 John H. Giltner, *Moses Stuart. The Father of Biblical Science in America* (Atlanta, 1988); Jay Gomer Williams, *Times and Life*, passim; Brown, *Rise of Biblical Criticism*, pp. 45–59 and passim.

5 Robinson to Tholuck, Andover, 28 September 1830, AFSt/ Thol: B III. 1120.

6 Robinson to Norton, Boston, 9 February 1837, ERP, ER 1.13. Cf. Smith and Hitchcock, *Life, Writings and Character of Robinson*, pp. 53–66; Jay Gomer Williams, *Times and Life*, pp. 94–111.

7 Robinson to Norton, Boston, 9 February 1837 (citation) and New York, 1 July 1837, ERP, ER 1.12; Bliss, *Development of Palestine Exploration*, pp. 188–223. On the founding and first years of the UTS, see Union Theological Seminary, *The Union Theological Seminary, in the City of New York, its History, Condition and Wants; an Appeal to the Friends of the Institution* (New York, 1852); Jay Gomer Williams, *Times and Life*, pp. 284–97.

8 Examples include: Bliss, *Development of Palestine Exploration*, pp. 184–223; Ben-Arieh, *Rediscovery*, pp. 85–91; Brown, *Rise of Biblical Criticism*, pp. 111–24; Robert T. Handy (ed.), *The Holy Land in American Protestant Life 1800–1948. A Documentary History* (New York, 1981), pp. 6–36 (but without using even one document); Silberman, *Digging for God and Country*, pp. 37–47; John Davis, *The Landscape of Belief. Encountering the Holy Land in Nineteenth-Century American Art and Culture* (Princeton, NJ, 1996); Michael B. Oren, *Power, Faith and Fantasy. America in the Middle East 1776 to the Present* (New York and London, 2007), pp. 134–9; Rachel

Hallote, *Bible, Map, and Spade. The American Palestine Exploration Society, Frederick Jones Bliss, and the Forgotten Story of Early American Biblical Archaeology* (Piscataway, NJ, 2006). But the best proof is the recently published book, Soler, *Edward Robinson*, where the author writes at the beginning of his bibliography: 'Des sources manuscrites existent, mais leur consultation nous a été impossible' (p. 291).

9 Jay Gomer Williams, *Times and Life*, pp. 113–74, citation p. 117.

10 Ibid., pp. 350–9.

11 Hermine Weigel Williams, *Robinson's Letter-Journal (1826–1829). Written from Europe by Edward Robinson to His Sister Elisabeth* (New York and Bloomington, 2009).

12 Robinson published a long obituary: Edward Robinson, 'Biographical notices of Gesenius and Nordheimer', *BS* 1/2, pp. 361–79. See also, for Stuart and Gesenius, Giltner, *Moses Stuart*, pp. 75–84.

13 Jay Gomer Williams, *Times and Life*, pp. 146–7.

14 Robinson to Norton, 1 July 1837, ERP, ER 1.12; Ulrich Jasper Seetzen, *Verzeichniss der für die orientalische Sammlung in Gotha zu Damask, Jerusalem u. s. w. angekauften orientalischen Manuscripte und gedruckten Werke, Kunst- und Naturprodukte u. s. w.* (ed. Ernst Friedrich Karl Rosenmüller, Leipzig, 1810). Rev. Asahel Strong Norton, brother of Edward Robinson's first wife's mother, was minister of the Congregational Church in Clinton, NY, and very close to Robinson ever since his studies at Hamilton College. For Seetzen, his travels, collections and publications: Friedrich Kruse et al. (eds), *Ulrich Jasper Seetzen's Reisen durch Syrien, Palästina, Phönicien, die Transjordan-Länder, Arabia Petraea und Unter-Aegypten* (4 vols, Berlin, 1854), I, pp. III–XXXIX; Yehoshua Ben-Arieh, 'Pioneer scientific exploration in the Holy Land at the beginning of the nineteenth century', *Terrae Incognitae* 4 (1972); Jutta Schienerl, *Der Weg in den Orient. Der Forscher Ulrich Jasper Seetzen. Von Jever in den Jemen (1802–1811)* (Oldenburg, 2000); Haim Goren, *Dead Sea Level. Science, Exploration and Imperial Interests in the Near East* (London, 2011), pp. 139–43; Dietmar Henze, *Enzyklopädie der Entdecker und Erforscher der Erde* (5 vols, Graz 1978–2004, repr. Darmstadt, 2011), V, pp. 124–9; Petra Weigel et al., *Das Heilige Land in Gotha. Der Verlag Justus Perthes und die Palästinakartographie im 19. Jahrhundert* (Gotha, 2014), pp. 34–9.

15 Annemarie Meiner, *Geschichte des Verlages Johann Ambrosius Barth Leipzig 1780–1930* (Leipzig, 1930); Robinson to Barth, Leipzig, 23 November 1838, Berlin, 13 December 1838, Berlin, 6 March 1840, ERP, ER 1.10. 2,500 volumes from Rosenmüller's library reached the University Library in Leipzig, most likely those dealing with theology and Oriental Sciences (https://www.ub.uni-leipzig.de/ueber-uns/geschichte/ zweite-periode-1833-1932/ accessed 18 March 2016). Smith's archive does include a handwritten document, 'titles of Arabic Books printed since 1810. Found in E.F.C. Rosenmüller's Library', ESP, ABC 60, (118) Lists of Books.

16 Robinson to Ritter, New York, 21 December 1840, SBB-PK, Nachl. Carl Ritter, K 5, p. 32; Robinson to Lippert, New York, 15 July 1841, SBB-PK, Darm. Asien, p. 5. Johann Friedrich Lippert, book dealer and auctioneer from Halle, bought for Robinson books sold at auction from the collection of the deceased professor Peter von Bohlen from Königsberg (Robinson to Lippert, 18 September 1840, ERP, ER 1.10).

17 [Joseph Daling] Ibbotson, 'Edward Robinson and his books: a biographical and descriptive sketch prepared for the Faculty Club in connection with an exhibit of books from the Robinson library' (1933), 13 leaves, typewritten, ERP, AL COLL I2 E3 1933: 'out of 42 in Röhricht [Röhricht, *Bibliotheca Geographica Palaestinae*, 1–20, H.G.] before 1095, they miss only 4'; Ibbotson, 'Edward Robinson and his books', print, signed 12 July 1952, ERP: 'Our entire Palestine collection numbers some 1200 volumes'.

18 Jay Gomer Williams, *Times and Life*, 130, 209–11.

19 Ibid., p. 192.

20 The organisation, largest and most important nineteenth-century American missionary society, has been well-researched. For example: Joseph L. Grabill, *Protestant Diplomacy and the Near East. Missionary Influence on American Policy, 1810–1927* (Minneapolis, 1971); Clifton Jackson Phillips, *Protestant America and the Pagan World. The First Half Century of the American Board of Commissioners for Foreign Missions, 1810–1860* (Cambridge, MA, 1969); Oren, *Power, Faith and Fantasy*, pp. 83–97, 122–34.

21 Eli Smith and Harrison Gray Otis Dwight, *Missionary Researches in Armenia Including a Journey through Asia Minor, and into Georgia and Persia, with a Visit to the Nestorian and Chaldean Christians of Oormiah and Salmas* (London, 1834); Margaret R. Leavy, 'Looking for the Armenians. Eli Smith's missionary adventure, 1830–1831', *Transactions of the Connecticut Academy of Arts and Science* 50 (1992); Grabill, *Protestant Diplomacy*, pp. 8–9. Interesting remarks concerning Smith's life and activity after his arrival in Beirut were written by a friend and colleague, William Goodell, who worked for forty years among the Armenians, probably as a necrologue, following Smith's death: Goodell to Thompson, Constantinople, 14 May 1859, ESP, ABC 60; cf. also Dagmar Glass, *Malta, Beirut, Leipzig and Beirut again. Eli Smith, the American Syria Mission and the Spread of Arabic Typography in 19th Century Lebanon* (Beirut, 1998).

22 Smith to Robinson, Smyrna, 10 November 1837, ESP, ABC 60 (105), Letters to Various Correspondents; Robinson and Smith, *Biblical Researches* (Boston), 2, cited by Jay Gomer Williams, *Times and Life*, p. 194.
23 Edward W. Hooker, *Memoir of Mrs. Sarah L. Huntington Smith, Late of the American Mission in Syria* (New York, 1846), pp. 174, 202, 238, 244, 312.
24 Edward Robinson and Eli Smith, *Biblical Researches in Palestine, Mount Sinai and Arabia Petraea in 1838. A Journal of Travels in the Year 1838*, I–III (Boston, 1841), vol. 3, p. 89.
25 Hooker, *Memoir of Mrs. Sarah L. Huntington Smith*, p. 231; 'Lists of the Arabic Names of Places in Palestine and the Adjacent Regions', Second Part, p. 139, ESP, ABC 60; published in Robinson and Smith, *Biblical Researches* (Boston), Second Appendix, pp. 89–171.
26 Examples of Eli Smith's own works: 'An ancient temple on Mt. Lebanon', *BS* 1 (1843) and 'The Wines of Mount Lebanon', *BSTR* 3 (1846).
27 Robinson to Washington, Berlin, 19 June 1839, RGS-IBG Collections, RGS Corr. Block 1834–40, RGS/CB2/454: Robinson; Robinson and Smith, *Biblical Researches* (Boston), I, p. 2.
28 Smith to Farren, Beirut, 5 and 9 May 1834, ESP, ABC 60 (105).
29 René Tresse, 'L'installation du premier consul d'Angleterre à Damas (1830–1834)', *Revue d'histoire des colonies* 24 (1936); cf. Judith Blow Williams, *British Commercial Policy and Trade Expansion 1750–1850* (Oxford, 1972), pp. 424–7.
30 Robinson and Smith, *Biblical Researches* (London), I, p. 2. Cf. Smith and Hitchcock, *Life, Writings and Character of Robinson*, pp. 74–5; Jay Gomer Williams, *Times and Life*, 219–21.
31 [Eli Smith], Rules of the orthography of Arabic and Turkish names and words, in Roman letters. Adopted by the Syrian Mission of the A.B.C.F.M., April 1838, ESP, ABC 60 (116); cf. Robinson and Smith, *Biblical Researches* (Boston), III, Second Appendix, pp. 89–111.
32 Smith and Hitchcock, *Life, Writings and Character of Robinson*, p. 74.
33 Robinson to Smith, Alexandria, 3 January 1838, ESP, ABC 60, Robinson, Edward.
34 Robinson to Smith, Cairo, 19 January 1838, ESP, ABC 60, Robinson, Edward. Schubert, a leading German natural historian from Munich whose work combined the influences of Romanticism with German *Naturphilosophie*, went on a long pilgrimage–exploration expedition in 1836–37 (Gotthilf Heinrich von Schubert, *Reise in das Morgenland in den Jahren 1836 und 1837*, I–III (Erlangen, 1839); Haim Goren, *'Zieht hin und erforscht das Land'. Die deutsche Palästinaforschung im 19. Jahrhundert* (Göttingen, 2003), pp. 117–24; Goren, *Dead Sea Level*, pp. 207–11). Lindsay, 'British nobleman, traveller, and writer on art', student at Trinity College, Dublin, spent his life in studious pursuits, in the collection of a magnificent library and in travel. In 1836–7 he travelled in Egypt and Palestine, his letters were published in 1838 (Alexander William Crawford, Lord Lindsay, *Letters on Egypt, Edom, and the Holy Land*, I–II (London, 1858)).
35 Smith to Robinson, Smyrna, 6 January 1838, ESP, ABC 60, Robinson, Edward.
36 Robinson to Smith, Berlin, 22 January 1839, ESP, ABC 60, Robinson, Edward.
37 Carl Ritter, *The Comparative Geography of Palestine and the Sinaitic Peninsula* (trans. William L. Gage) (4 vols, New York, 1866, repr. 1968), II, p. 74.
38 Robinson to Smith, Berlin, 10 January 1839, ESP, ABC 60, Robinson, Edward. And see Glass, *Malta, Beirut, Leipzig*.
39 Robinson to Smith, Constantinople, 7 August 1838, ESP, ABC 60, Robinson, Edward.
40 Theresa Robinson to Smith, Dresden, 21 September 1838, ESP, ABC 60, Robinson, Edward.
41 Robinson to Smith, Vienna, 24 October 1838, Halle, 14 November 1838, ESP, ABC 60, Robinson, Edward.
42 Smith to Anderson, Smyrna, 4 August 1838, ESP, ABC 60 (105). Rufus Anderson studied in Andover and from 1832 was in charge of all overseas work of the ABCFM.
43 Smith to Robinson, Leipzig, 8 February 1839, ESP, ABC 60 (105).
44 Robinson to Smith, Halle, 14 November 1838, ESP, ABC 60, Robinson, Edward.
45 Robinson to Smith, Berlin, 22 January 1839, ESP, ABC 60, Robinson, Edward.
46 Ibid.
47 Robinson to Smith, Berlin, 23 November 1838, ESP, ABC 60, Robinson, Edward. The presented lecture had been published: Edward Robinson and Eli Smith, 'Kurzer Bericht über eine Reise nach Palästina und der Umgegend im Jahr 1838, in Bezug auf die biblische Geographie unternommen', *Zeitschrift für die Kunde des Morgenlandes* 2 (1839).
48 Robinson to the RGS, 18 June 1939, RGS-IBG Collections, JMS/9/49; Edward Robinson, 'Extracts from a journal of travels in Palestine &c., in 1838; undertaken for the illustration of Biblical Geography', *JRGS* 9 (1839).
49 Prudhoe to Smith, Jerusalem, 15 April 1838, ESP, ABC 60, General letters N-O-P: Lord Prudhoe.

50 John Ruffle, 'The Journeys of Lord Prudhoe and Major Orlando Felix in Egypt, Nubia and the Levant 1826–1829', in P. Starkey and J. Starkey (eds), *Travellers in Egypt* (London and New York, 2001); Warren R. Dawson and Eric P. Uphill, *Who Was Who in Egyptology*, 3rd rev. edn (ed. M. L. Bierbrier, London, 1995), p. 325.

51 Robinson to Smith, Berlin, 24 January 1852, ESP, ABC 60, Robinson, Edward. Syon House and park, close to Kew Gardens, still belong to the Duke of Northumberland.

52 Lanneau to Robinson and Smith, Jerusalem, 5 June 1838, ESP, ABC 60, Lanneau, John F. ABCFM missionary John Francis Lanneau of South Carolina served in Palestine from 1836 to 1839.

53 Whiting to Smith, Jerusalem, 4 June 1838, ESP, ABC 60, Whiting, George B. Tibawi, Abdul Latif, *American Interests in Syria 1800–1901: A Study of Educational, Literary and Religious Work* (Oxford, 1966), pp. 57–62, 74, 111, 130; Kawerau, Peter, *Amerika und die orientalischen Kirchen: Ursprung und Anfang der amerikanischen Mission unter den Nationalkirchen Westasiens* (Berlin, 1958), pp. 254–7. For Whiting's biographical details: Personnel records for George B. Whiting, American Board Personnel Card and Photo Collection, Digital Library for International Research (accessed 1 July 2015).

54 Robinson to Smith, Berlin, 22 January 1839, ESP, ABC 60, Robinson, Edward.

55 Robinson to Smith, Berlin, 29 March 1839, ESP, ABC 60, Robinson, Edward.

56 Robinson to Smith, Berlin, 17 April 1839, ESP, ABC 60, Robinson, Edward.

57 Robinson to Smith, Berlin, 24 December 1839, ESP, ABC 60, Robinson, Edward.

58 Robinson to Smith, Berlin, 12 February 1840, ESP, ABC 60, Robinson, Edward.

59 Robinson to Smith, Berlin, 15 April 1840, ESP, ABC 60, Robinson, Edward.

60 Robinson to Ritter, New York, 15 March 1840, SBB-PK, Nachl. Carl Ritter, K 5, p. 30.

61 Robinson to Ritter, New York, 21 December 1840, SBB-PK, Nachl. Carl Ritter, K 5, p. 32.

62 Robinson and Smith, *Biblical Researches* (London) and *Biblical Researches* (Boston); Edward Robinson and Eli Smith, *Palästina und die südlich angrenzenden Länder. Tagebuch einer Reise im Jahre 1838 in Bezug auf die biblische Geographie unternommen*, I–III (Halle, 1842).

63 Robert Cooke (for his uncle, John Murray) to Robinson, London, 3 August 1841, ERP, L. Correspondence to E.R. 1840–60; Robinson to Ritter, New York, 1 October 1841, SBB-PK, Darm. Asien, p. 6; Robinson to Ritter, New York, 31 December 1841, SBB-PK, Nachl. Carl Ritter, K 5, p. 34.

64 Robinson to Ritter, New York, 27 February 1843, SBB-PK, Nachl. Carl Ritter, K 5, p. 36; Robinson to Smith, New York, 18 February 1843, ESP, ABC 60, Robinson, Edward.

65 Robinson to Ritter, New York, 30 December 1843, SBB-PK, Nachl. Carl Ritter, K 5, p. 38. 'Mammon' meaning wealth, richness, as in Hebrew.

66 Robinson to Ritter, Halle, 30 October 1852, UBF, Ritter NL 33/144/4. Georg was the publisher and his brother Dietrich the assortments dealer and literary person (Ritter to Robinson, Berlin, 25 September 1852, ERP, L 4 Ritter).

67 Robinson to Ritter, Bremen, 7 October 1852, UBF, Ritter NL 33/144/5; Robinson to Smith, New York, 11 December 1852, ESP, ABC 60, Robinson, Edward.

68 Murray to Robinson, 13 October 1853, ERP, L. Correspondence to E.R. 1840–60.

69 Robinson to Smith, New York, 20 March 1855, ESP, ABC 60, Robinson, Edward.

70 Robinson to Ritter, New York, 24 March 1855, UBF, Ritter NL 33/144/7.

71 Murray to Robinson, London, 7 July 1856, ERP, L. Correspondence to E.R. 1840–60.

72 Ibid.

73 Murray to Ritter, 7 July 1856, ERP, L. Correspondence to E.R. 1840–60.

74 Robinson to Ritter, New York, 15 March 1840, SBB-PK, Nachl. Carl Ritter, K 5, p. 30.

75 Robinson to Ritter, New York, 31 December 1840, SBB-PK, Nachl. Carl Ritter, K 5, p. 33.

76 Smith and Hitchcock, *Life, Writings and Character of Robinson*, pp. 659–70; Goren, *Zieht hin*, p. 85–91; Goren, 'Carl Ritter's Contribution'.

77 Carl Ritter, *Die Erdkunde im Verhältnis zur Natur und zur Geschichte des Menschen oder allgemeine vergleichende Geographie als sichere Grundlage des Studiums und Unterrichts in physicalischen und historischen Wissenschaften* (Berlin, 1818–1858), 2nd part, 3rd book: *Westasien*, 3rd unit: *Das Plateau von Soristan, oder das syrische Gebirgsland*, pp. 299–467.

78 Carl Ritter, *Vergleichende Erdkunde der Sinai-Halbinsel, von Palästina und Syrien*, I–IV (Die Erdkunde im Verhältniß zur Natur und zur Geschichte des Menschen, oder allgemeine vergleichende Geographie, XIV-XVII, Berlin, 1848–55); Ritter, *Comparative Geography*. Cf. Goren, 'Carl Ritter's Contribution'; Goren, *Zieht hin*, pp. 68–83.

79 Otto Richter, *Der teleologische Zug im Denken Carl Ritters* (Borna und Leipzig, 1905); David N. Livingstone, *The Geographical Tradition. Episodes in the History of a Contested Enterprise* (Oxford, Cambridge, 1992), pp. 139–41;

Goren, *Zieht hin*, pp. 71–3. Cf. also various contributions in Hans Richter (ed.), *Carl Ritter. Werk und Wirkungen* (Gotha, 1983).

80 ERP, M.2 ER's letter-journal 1826–29; Jay Gomer Williams, *Times and Life*, p. 162. Cf. Smith and Hitchcock, *Life, Writings and Character of Robinson*, p. 69.

81 See Jay Gomer Williams's detailed discussion concerning Stuart's directions and instructions: *Times and Life*, passim; citation: ibid., p. 260. Cf. also Giltner, *Moses Stuart*, pp. 75–84.

82 Jay Gomer Williams, *Times and Life*, p. 309; Robinson to Ritter, New York, 30 December 1843, SBB-PK, Nachl. Carl Ritter, K 5, p. 38.

83 Jay Gomer Williams, *Times and Life*, p. 261.

84 Gotthilf Heinrich von Schubert, 'Palästina und die südlich angränzenden Länder. Tagebuch einer Reise im Jahre 1838 ... von E. Robinson und E. Smith ... Halle 1841', *Gelehrte Anzeigen* 13 (1841), p. 681.

85 *The Quarterly Review*, 69/1 (December 1841), pp. 78–97, citation p. 78. Cf. Jay Gomer Williams, *Times and Life*, p. 359, for 'selected reviews'.

86 Smith and Hitchcock, *Life, Writings and Character of Robinson*, p. 89.

87 cf. Yehoshua Ben-Arieh, 'Perceptions and images of the Holy Land', in R. Kark (ed.), *The Land that Became Israel. Studies in Historical Geography* (New York, London, Jerusalem, 1989); Yehoshua Ben-Arieh, 'Holy Land views in 19th century western travel literature', in M. Davis and Y. Ben-Arieh (eds), *Western Societies in the Holy Land* (Westport, 1991); Goren, 'Carl Ritter's Contribution'.

88 Edward Robinson, *Physical Geography of the Holy Land* (London, 1865); cf. Bliss, *Development of Palestine Exploration*, pp. 221–2; Ben-Arieh, *Rediscovery*, p. 154.

89 George Adam Smith, *The Historical Geography of the Holy Land* (London, 1974 [1894]), p. 90. Cf. Robin Butlin, 'George Adam Smith and the historical geography of the Holy Land. Contents, contexts and connections', *Journal of Historical Geography* 14/4 (1988); Edwin James Aiken, *Scriptural Geography. Portraying the Holy Land* (London, 2010), pp. 18–56, 133–85; Ben-Arieh, 'Perceptions and images', pp. 37–45.

90 Carl Ritter, 'Neue Entdeckungen über die Jordan-Quelle von Major Robe, und über einzelne Punkte von Palästina durch Wolcott, nach Mittheilungen von Herrn Robinson', *MVGEB* 5 (1842–3), p. 126.

91 Review in the *Wiener Jahrbuch der Literatur*, 1842 and 1843. Justus Olshausen, professor of Oriental languages, expert in Semitic and Persian philology.

92 Ritter, *Comparative Geography*, II, pp. 70–1; Carl Ritter, *Die Erdkunde von Asien*. Vol. VIII, second unit: *Die Sinai-Halbinsel, Palästina und Syrien*. Second section, first unit: *Palästina und Syrien* (Berlin, 1850), pp. 85–8.

93 Ibid., II, p. 71. He mentions here some French critics, as well as George Williams in his *Holy City* published in 1845, in his words 'prelatical England and Catholic France' (George Williams, *The Holy City. Historical, Topographical, and Antiquarian Notices on Jerusalem*, I–II (London, 1849), p. 72).

94 Ritter, *Comparative Geography*, II, pp. 71–2.

95 Robinson to Smith, Halle, 14 November 1838, ESP, ABC 60, Robinson, Edward.

96 Robinson to Smith, Halle, 10 September 1839, ESP, ABC 60, Robinson, Edward.

97 Eißfeldt, 'Wilhelm Gesenius'. Cf. Smith and Hitchcock, *Life, Writings and Character of Robinson*, pp. 48–52; Jay Gomer Williams, *Times and Life*, pp. 138–45 and passim.

98 Jay Gomer Williams, *Times and Life*, pp. 137–42 and passim.

99 Ritter, *Comparative Geography*, II, pp. 72–3.

100 Robinson to Smith, Berlin, 11 February 1839, ESP, ABC 60, Robinson, Edward. Cf. Grimm's handwritten memoir, SBB-PK, Ms. Germ. Quart. 1056, pp. 12–24; the map: Julius Ludwig Grimm, *Palaestina, 1:900.000* (Berlin, 1830) (see Fig. 1.2); cf. Haim Goren, 'Carl Ritter ermutigt junge Kartographen', in P. Weigel et al. (eds), *Das Heilige Land in Gotha. Der Verlag Justus Perthes und die Palästinakartographie im 19. Jahrhundert* (Gotha, 2014). It is a bit strange and unjustified, that Robinson writes about Grimm's 'French maps'.

101 Ritter to Robinson, Berlin, 18 March 1850, ERP, L.4 Ritter.

102 Ritter to Robinson, Berlin, 6 May 1851; 29 June 1851, ERP, L.4 Ritter wrote in German; underline in original.

103 Charles William Meredith van de Velde, *Memoir to Accompany the Map of the Holy Land* (Gotha, 1858), p. 4. See Chapter 2 in this volume.

104 Both Washington and Mahlmann are discussed in this chapter.

105 Kiepert to Ritter, Berlin, 21 December 1843, Nachl. 133 (Carl Ritter), K 3, p. 33. For Robe's and his map cf. Haim Goren, 'Pioneer surveying and mapping of the Galilee and the Hauran. The Map of the Sources of the Jordan River (1841) and Map of the *Ledja* (1838)', *Cathedra* 96 (July 2000). See also Wolcott's letters: to Smith, Jerusalem, 30 December 1841 and to Robinson and Smith, Belchestown, Mass., 1 May 1852, ESP, ABC 60, Wolcott, Samuel.

106 Ritter, 'Neue Entdeckungen'; Edward Robinson, 'Researches in Palestine. Compiled by the editor from various communications received at different times from the Rev. Eli Smith and the Rev. S. Wolcott. With a Map of the Sources of the Jordan', *BS*, 1/1 (1843), pp. 9–15; Yolande Jones, 'British military surveys of Palestine and Syria 1840–1841', *The Cartographic Journal* 10/1 (1973); Goren, 'Pioneer Surveying'; Haim Goren, 'British surveyors in Palestine and Syria, 1840–1841', International Cartographic Conference A Coruña 2005: *Mapping Approaches into a Changing World*, available at http://icaci.org/documents/ICC_proceedings/ICC2005/htm/pdf/oral/TEMA16/Session%205/HAIM%20GOREN.pdf (accessed 13 August 2016).
107 Robinson, 'Researches in Palestine', pp. 17, 41–6.
108 Samuel Wolcott, 'Geographical notes on Palestine. The coast of the Dead Sea, Masada, route from Mount Lebanon to Baalbek', *BS* 3 (1846); Samuel Wolcott, 'The topography of Jerusalem', *BS* 23 (1866).
109 Kiepert to Ritter, New York, 15 November 1843, Nachl. 133 (Carl Ritter), K 3, p. 42; Robinson to Smith, New York, 11 December 1845, ESP, ABC 60, Robinson, Edward. Cf. George Williams, *Holy City*; Bliss, *Development of Palestine Exploration*, pp. 232–3.
110 Kiepert to Ritter, New York, 15 November 1843, Nachl. 133 (Carl Ritter), K 3, p. 42. Cf. Ernst Gustav Schultz, *Jerusalem. Eine Vorlesung, mit einem Plane gezeichnet von H. Kiepert* (Berlin, 1845); Goren, *Zieht hin*, pp. 194–201.
111 Robinson to Ritter, Stralsund (in Pomerania), 8 July 1851, UBF, Ritter NL 33/144/2. It is interesting to find out, how these relative late letter exchange between both 'landed' in a collection at Freiburg.
112 Robinson to Ritter, Halle, 31 July 1851, UBF, Ritter NL 33/144/3.
113 Washington to Robinson, Woolwich, 10 January 1852, ERP, L. Correspondence to E.R. 1840–60.
114 Ritter to Robinson, Berlin, 29 June 1851, ERP, L.4 Ritter.
115 Edgar Lehmann, 'Carl Ritters kartographische Leistung', *Die Erde* 90 (1959).
116 Heinrich Berghaus, *Geographisches Memoir zur Erklärung und Erläuterung der Karte von Syrien*, Vol. I (Gotha, 1835); Ritter, *Comparative Geography*, Vol. 2, pp. 80–1; Gerhard Engelmann, *Heinrich Berghaus. Der Kartograph von Potsdam* (Halle/Saale, 1977), pp. 63, 107, 118–19; Heinrich Kiepert, 'Memoir on the maps accompanying this work', in Robinson and Smith, *Biblical Researches* (London, 1841); Hans Fischer, 'Geschichte der Kartographie von Palästina', *ZDPV* 62 (1939), pp. 169–89 and 63 (1940), pp. 43–52; Petra Weigel, 'Heinrich Berghaus verknüpft Zahlen, Karten und Berichte', in P. Weigel et al. (eds), *Das Heilige Land in Gotha. Der Verlag Justus Perthes und die Palästinakartographie im 19. Jahrhundert* (Gotha, 2014).
117 See for example SBB-PK, Nachl. Carl Ritter, K 1, pp. 78, 79, concerning maps of Kurdistan and Africa.
118 Joseph Partsch, 'Heinrich Kiepert. Ein Bild seines Lebens und seiner Arbeit', *Geographische Zeitschrift* 7 (1901), pp. 4–5, 11–16, 77–88.
119 Jay Gomer Williams, *Times and Life*, p. 262.
120 Berghaus to Washington, Potsdam, 30 April 1839, RGS-IBG Collections, Corr Block 1834–40, RGS/CB2. The German version: Berghaus, 'Über die Reisen', p. 531.
121 John Lewis Burckhardt, *Travels in Syria and the Holy Land* (London, 1822); Ben-Arieh, 'Pioneer scientific exploration', pp. 103–6; Goren, *Zieht hin*, pp. 48–55 and the bibliography there.
122 Robinson to Smith, Berlin, 22 January 1839, HLH, ABC 60, Robinson, Edward.
123 Engelmann, *Heinrich Berghaus*. Cf. also Goren, *Zieht hin*, pp. 92–5 for his Holy Land cartography, before consulting the archives.
124 Gerhard Engelmann, 'Der Physikalische Atlas des Heinrich Berghaus und Alexander Keith Johnstons Physical Atlas', *PGM* 108 (1964); Gerhard Engelmann, 'August Petermann als Kartographenlehrling bei Heinrich Berghaus in Potsdam.', *PGM* 106 (1962).
125 Ritter, *Comparative Geography*, II, pp. 80–1 (citation); van de Velde, *Memoir*, pp. 15–16; Fischer, 'Geschichte der Kartographie', pp. 43–5 (detailed list of Berghaus's sources).
126 Van de Velde, *Memoir*, p. 16.
127 Heinrich Berghaus, *Karte von Syrien, den Manen Jacotin's und Burckhardt's gewidmet* (Gotha, 1835) (see Plate 1). Charles W. Wilson, 'Recent surveys in Sinai and Palestine', *JRGS* 43 (1873), p. 212; Fischer, 'Geschichte der Kartographie', pp. 43–5; Weigel, 'Heinrich Berghaus'. For Jacotin and his maps (Pierre Jacotin, *Carte topographique de l`Égypte et de plusieurs parties des pays limitrophes, levée pendant l'expédition de l'armée française*, 47 sheets, Paris, 1810)), cf. Berghaus, *Geographisches Memoir*, Vol. I, pp. 1–4; Yehuda Karmon, 'An analysis of Jacotin's Map of Palestine', *Israel Exploration Journal*, 10/3 (1960); Anne Godlewska, *Geography Unbound. French Geographical Science from Cassini to Humboldt* (Chicago, 1999), mainly pp. 77–8; Anne Godlewska, 'Napoleon's geographers (1797–1815). Imperialists and soldiers of modernity', in A. Godlewska and N. Smith (eds), *Geography and Empire* (Oxford and Cambridge, MA, 1994); Anne Godlewska, *The Napoleonic Survey of Egypt. A Masterpiece of Cartographic Compilation and Early Nineteenth-Century Fieldwork* (Toronto, 1988), passim.

128 Robinson to Smith, Berlin, 22 January 1839, ESP, ABC 60, Robinson, Edward.
129 Robinson to Smith, Berlin, 24 December 1839, ESP, ABC 60, Robinson, Edward.
130 Ibid.
131 Edward Robinson, 'Extracts from a Journal'.
132 The history of the publication of this map has been published as a paper: Haim Goren and Bruno Schelhaas, 'Berghaus's Part of Arabia Petraea and Palestine Map (1839) and the Royal Geographical Society. Misuse or misunderstanding?', *Terrae Incognitae*, 47(2). Described here are details relevant to the current paper.
133 Jules de Bertou, 'Notes on a journey from Jerusalem to Hebron, the Dead Sea, El Ghór, and Wádi 'Arabah to 'Akabah, and back by Petra, in April, 1838', *JRGS* 9 (1839); Jules de Bertou, 'Extract from a notice on the site of ancient Tyre', *JRGS* 9 (1839). See also Goren, *Dead Sea Level*, esp. pp. 211–8 and the French wikipedia article about Bertou https://fr.wikipedia.org/wiki/Jules_de_Bertou (accessed 7 December 2016).
134 Robinson to Smith, Berlin, 10 July 1839, ESP, ABC 60, Robinson, Edward.
135 See in detail Goren and Schelhaas, 'Berghaus's Part of Arabia Petraea and Palestine Map (1839)'.
136 Washington to Robinson, Shearwater, Harwich, 1 May 1842, ERP, L.2 Letters from J. Washington (citation); Robinson to Ritter, 2 November 1840, SBB-PK, Nachl. Carl Ritter, K 5, p. 31; Dawson, *Memoirs of Hydrography*, pp. 93–111; Archibald Day, *The Admiralty Hydrographic Service, 1795–1919* (London, 1967), pp. 67–80; J. K. Laughton (rev. R. O. Morris), 'Washington, John (1800–1863)', in *ODNB*, 101028807 (2004); Ritchie, *Admiralty Chart*, pp. 281–3.
137 Washington to Robinson, Shearwater, Harwich, 1 May 1842, ERP, L.2 Letters from J. Washington.
138 Washington to Berghaus, London, 1 September 1837, RGS-IBG Collections, Letter-Book 1836–40.
139 Ritter's copy of Beaufort's letter to anonymous, 4 April 1842, SBB-PK, Nachl. Carl Ritter, K 1, p. 58.
140 Robinson to Smith, Berlin, 10 January 1839, ESP, ABC 60, Robinson, Edward.
141 Robinson to Smith, Berlin, 24 December 1839, ESP, ABC 60, Robinson, Edward.
142 Berghaus to Washington, Potsdam, 30 April 1839, RGS-IBG Collections, Corr Block 1834–40, RGS/CB2.
143 Robinson to Smith, Berlin, 22 January 1839, ESP, ABC 60, Robinson, Edward. Cf. Robinson and Smith, *Biblical Researches* (Boston), III, First Appendix, C. Itinerary.
144 Robinson to Smith, Berlin, 14 March 1839, ESP, ABC 60, Robinson, Edward.
145 Heinrich Berghaus, 'Über die Reisen von E. Robinson und E. Smith im Peträischen Arabien und im Gelobten Lande', *Almanach für das Jahr 1840. Der Belehrung und Unterhaltung auf dem Gebiete der Erd-, Länder-, Völker- und Staatenkunde* 4 (*1840*), p. 531; emphasis in original.
146 Robinson to Washington, Berlin, 4 June 1839, RGS-IBG Collections, RGS Corr. Block 1834–40, RGS/CB2/454: Robinson.
147 Berghaus to Washington, Potsdam, 30 April 1839, RGS-IBG Collections, RGS Corr. Block 1834–40, RGS/CB2: Berghaus, Prof. For the travellers mentioned cf. Goren, *Zieht hin.*
148 Smith to Robinson, London, 8 July 1839, ESP, ABC 60 (105).
149 Robinson to Washington, Berlin, 18 September 1839, RGS-IBG Collections, RGS Corr. Block 1834–40, RGS/CB2/454: Robinson.
150 Engelmann, 'August Petermann', p. 173.
151 Washington to Berghaus, London, 5 July 1839, RGS-IBG Archive, RGS Letter Book 1836–40, p. 272.
152 Berghaus to Washington, Potsdam, 30 April 1839, RGS-IBG Collections, RGS Corr. Block 1834–40, RGS/CB2: Berghaus, Prof.
153 Washington to Berghaus, London, 5 July 1839, RGS-IBG Archive, RGS Letter Book 1836–40, pp. 272–6.
154 Robinson to Smith, Berlin, 24 December 1839, ESP, ABC 60, Robinson, Edward.
155 Karl Lenz, 'Heinrich Kiepert. Seine Stellung und Bedeutung in der Geographie', in L. Zögner (ed.), *Antike Welten. Neue Regionen. Heinrich Kiepert 1818–1899* (Berlin, 1999), pp. 11–13; Ulrich Freitag, 'Heinrich Kieperts kartographische Leistung', in L. Zögner (ed.), *Antike Welten. Neue Regionen. Heinrich Kiepert 1818–1899* (Berlin, 1999), passim.
156 Lothar Zögner (ed.), *Antike Welten. Neue Regionen. Heinrich Kiepert 1818–1899* (Berlin, 1999), pp. 79–81; cf. Haim Goren, 'Heinrich Kiepert in the Holy Land, spring 1870. Sketches from an exploration-tour of an historical cartographer', in L. Zögner (ed.), *Antike Welten. Neue Regionen. Heinrich Kiepert 1818–1899* (Berlin, 1999).
157 Robinson to Smith, Berlin, 24 December 1839, ESP, ABC 60, Robinson, Edward.
158 Robinson to Smith, Berlin, 12 February 1840, ESP, ABC 60, Robinson, Edward.
159 Robinson to Smith, Berlin, 15 April 1840, ESP, ABC 60, Robinson, Edward.

160 Frederic Catherwood was the first to measure and map the interior of the Temple Mount; cf. Yehoshua Ben-Arieh, 'The first surveyed maps of Jerusalem', *Eretz-Israel. Archaeological, Historical and Geographical Studies* 11 (1973) [Hebrew]. Robinson's archive holds a copy of the map (ERP, ABC 60).
161 Robinson to Smith, London, 3 October 1840, ESP, ABC 60, Robinson, Edward.
162 Kiepert, 'Memoir', pp. 29–30.
163 Robinson to Smith, London, 3 October 1840, ESP, ABC 60, Robinson, Edward: 'In Mount Lebanon we had not only Mr. Bird's map, but a very beautiful measuring map of Prof. Ehrenberg, which gives the *terrain* much better than Mr. Birds'.
164 Isaac Bird, *Bible Work in Bible Lands, or, Events in the History of the Syria Mission* (Philadelphia, 1872); cf. Makdisi, *Artillery of Heaven*, passim.
165 Ritter, *Comparative Geography,* II, pp. 82–3.
166 Ernst Plewe (ed.), *Die Carl Ritter Bibliothek* (Wiesbaden, 1978); cf. Lehmann, 'Carl Ritters kartographische Leistung'. Weigel inherited the publishing and auction house, established in 1795 by his father, in 1839.
167 For Ritter's estate in the Berlin State Library (*Inhaltsverzeichnis des Findbuchs Nachl. Carl Ritter*, SBB-PK): http://kalliope-verbund.info/DE-611-BF-26361 (accessed 8 November 2016). See also Gerhard Engelmann and Egon Klemp, 'Carl Ritters Briefsammlung in der Handschriftenabteilung/ Literaturarchiv der Deutschen Staatsbibliothek', in H. Richter (ed.), *Carl Ritter. Werk und Wirkungen* (Gotha, 1983); Plewe, *Carl Ritter Bibliothek*; Lothar Zögner, *Carl Ritter in seiner Zeit 1779–1859* (Berlin, 1979).
168 Plewe, *Carl Ritter Bibliothek*, pp. 206–13.
169 Ibid., pp. 59(477)–66(484).
170 Carl Sandreczki, 'Zur alten Geographie Palästina's', *Das Ausland* 5 (1872), p. 103. For Sandreczki (the father, not to be misidentified as his son, the physician Max S.), see Goren, *Zieht hin*, pp. 299–303.
171 Partsch, 'Heinrich Kiepert', pp. 11–12.
172 Goren, 'Carl Ritter's Contribution'; Goren, *Zieht hin*, pp. 68–83; Goren, 'Heinrich Kiepert in the Holy Land'.
173 Various comments in Kiepert's 'Memoir', as in pp. 34, 38. Captain Robert Moresby, hydrographer and maritime surveyor in the Indian Navy, pioneer surveyor of the Red Sea; cf. Henze, *Enzyklopädie der Entdecker*, 3, pp. 532–5.
174 Reinhold Röhricht, *Bibliotheca Geographica Palaestinae. Chronologisches Verzeichnis der auf die Geographie des Heiligen Landes bezüglichen Literatur von 333 bis 1878 und Versuch einer Cartographie* (Berlin, 1890, repr. Jerusalem, 1963, ed. David H.K. Amiran), pp. 638–9; for Robinson's publications see pp. 391–3.
175 Zögner, *Antike Welten*. See Appendix 1.
176 Tobler, *Bibliographia*, p. 239. For Tobler: Goren, *Zieht hin,* pp. 222–43; Hinderling and Sonderegger, 'Titus Tobler'.
177 Ritter, *Comparative Geography*, II, p. 70; cf. Kiepert, 'Memoir', p. 30, for his praising of Mahlmann's work 'with great fidelity, accuracy, and elegance'.
178 Ritter, *Comparative Geography*, II, p. 82.
179 Wolcott, 'Geographical Notes', p. 398.
180 Heinrich Kiepert, 'Topographische Beobachtung und Zeichnung', in G. von Neumayer (ed.), *Anleitung zu wissenschaftlichen Beobachtungen auf Reisen* (Berlin, 1875).
181 Jehuda L. Wallach, *Anatomie einer Militärhilfe. Die preußisch-deutschen Militärmission in der Türkei 1835–1919* (Düsseldorf, 1976); Friedrich Herre, 'Ein Preusse in der Türkei. Helmuth von Moltkes Mission im Osmanischen Reich', *Damals* 19/2 (1987). Cf. Lenz, 'Heinrich Kiepert', pp. 11–16.
182 Jay Gomer Williams, *Times and Life*, p. 261; Smith and Hitchcock, *Life, Writings and Character of Robinson*, p. 69.
183 Robinson to Ritter, Berlin, 22 January 1839, ESP, ABC 60, Robinson, Edward.
184 Robinson to Ritter, London, 2 November 1840, SBB-PK, Nachl. Carl Ritter, K 5, p. 31.
185 Carl Kassner, 'C. H. W. Mahlmann zum 125. Geburtstage', *Zeitschrift für angewandte Meteorologie* 53 (1936). See Alexander von Humboldt, *Central-Asien. Untersuchungen über die Gebirgsketten und die vergleichende Klimatologie* (2 vols, Berlin, 1844); Wilhelm Mahlmann, 'Bericht über die Witterungsbeobachtungen der Herrn Gen.-Consul v. Wildenbruch, Dr. van Dyck, Dr. de Forest und Hurter in Palästina, nach den von Herrn v. Wildenbruch eingesandten Beobachtungsjournalen', *MVGEB* N.F. 1 (1844).
186 Robinson to Ritter, London, 2 November 1840, SBB-PK, Nachl. Carl Ritter, K 5, p. 31.
187 Robinson to Kiepert, London, 2 November 1840, SBB-PK, Darm. Asien 1841: Robinson, Edward, pp. 3–4.
188 Johann Heinrich Westphal, 'Ueber die topographische Lage Jerusalems, zur Erläuterung des Planes aufgenommen von dem Herrn Dr. Westphal in Göttingen. Aus dessen Tagebuche während einer Reise durch

den Orient in den Jahren 1822 und 1823', *Hertha* 1/3 (1825); Haim Goren and Bruno Schelhaas, 'On an unknown measuring of Jerusalem in 1823', *Die Erde* 146/1 (2015).

189 Robinson to Smith, Berlin, 17 April 1839, ESP, ABC 60, Robinson, Edward.

190 Robinson to Smith, Berlin, 22 January 1839, ESP, ABC 60, Robinson, Edward. Robinson made two mistakes: he called him Parthei and noted the year of the measuring in Jerusalem as 1825 instead of 1823.

191 Karl von Raumer, 'Das östliche Palästina und das Land Edom. Zweiter Beitrag zur biblischen Erdkunde', *Annalen der Erd-, Völker- und Staatenkunde* 1 (1829–30); Goren, *Zieht hin*, pp. 144–7; Petra Weigel, 'Karl Georg von Raumer betreibt Bibelkartographie', in P. Weigel et al. (eds), *Das Heilige Land in Gotha. Der Verlag Justus Perthes und die Palästinakartographie im 19. Jahrhundert* (Gotha, 2014).

192 Karl von Raumer, *Der Zug der Israeliten aus Aegypten nach Canaan. Ein Versuch* (Leipzig, 1837); Raumer, *Palästina*.

193 Robinson to Smith, Berlin, 22 January 1839, ESP, ABC 60, Robinson, Edward.

194 Robinson to Kiepert, London, 2 November 1840, SBB-PK, Darm. Asien 1841: Robinson, Edward, pp. 3–4.

195 Malcolm Wagstaff, 'Colonel Leake. Traveler and scholar', in S. Searight and M. Wagstaff (eds), *Travellers in the Levant* (Durham, 2001); Wagstaff, 'Leake, William Martin (1777–1860)', in *ODNB*, 16242; James Marshall-Cornwall, 'Three Soldier-Geographers', *Geographical Journal* 131/3 (1965), pp. 357–9.

196 Leake to Ritter, London, 24 September 1841, SBB-PK, Nachl. Carl Ritter, K 3, p. 117.

197 For example: Wilhelm Friedrich Hemprich and Christian Gottfried Ehrenberg, *Naturgeschichtliche Reisen durch Nord-Afrika und West-Asien in den Jahren 1820 bis 1825* (Berlin, Posen, Bromberg, 1828); Christian Gottfried Ehrenberg, *Die Corallenthiere des rothen Meeres, physiologisch untersucht und systematisch verzeichnet* (Berlin, 1834). Cf. Alexander von Humboldt, 'Rapport sur le voyage fait par MM. Ehrenberg et Hemprich en Égypte, Dongola, Syrie, Arabie, et à la pente orientale du plateau de l'Abyssinie, de 1820 à 1824', *Nouvelles Annales des Voyages et des Sciences Géographiques* 36 (1827); Erwin Stresemann, *Hemprich und Ehrenberg. Reisen zweier naturforschender Freunde im Orient, geschildert in ihren Briefen aus den Jahren 1819–1826* (Berlin, 1954); Goren, *Zieht hin*, 60–2.

198 For example 'Karte des Nils und des Golfes von Suez', Hemprich and Ehrenberg, *Naturgeschichtliche Reisen*, vol. I, part I.

199 Cf. Bruno Schelhaas's contribution, Chapter 3 in this volume; Heinz Peter Brogiato, 'Gotha als Wissens-Raum', in S. Lentz and F. Ormeling (eds), *Die Verräumlichung des Welt-Bildes. Petermanns Geographische Mitteilungen zwischen 'explorativer Geographie' und der 'Vermessenheit' europäischer Raumphantasien* (Stuttgart, 2008).

200 Robinson to Ritter, London, 2 November 1840, SBB-PK, Nachl. Carl Ritter, K 5, p. 31; Robinson to Kiepert, London, 2 November 1840, SBB-PK, Darm. Asien 1841: Robinson, Edward, pp. 3–4.

201 Ibid.

202 Robinson to Ritter, New York, 31 December [1840], SBB-PK, Nachl. Carl Ritter, K 5, p. 33.

203 Robinson to Ritter, London, 2 November 1840, SBB-PK, Nachl. Carl Ritter, K 5, p. 31.

204 Robinson to Ritter, London, 21 December 1840, SBB-PK, Nachl. Carl Ritter, K 5, p. 32; and see cited letter in n. 96 above.

205 Ibid. Hermann Agathon Niemeyer managed the Francke Foundations (Frankeschen Stiftungen) in Halle, including the publishing house, in the years 1829–51.

206 James Silk Buckingham, *Travels in Palestine. Through the Countries of Bashan and Gilead, East of the River Jordan, Including a Visit to the Cities of Geraza and Gamala in the Decapolis* (2 vols, London, 1822); Smith to 'Messrs. S. E. Morse & es. Eds N. Y. Observer', Rochester, 28 January 1839; also numerous letters of Robinson to Smith, 1839–40, ESP, ABC 60; Bird to Smith, Gilmanton, 16 March 1840, ESP, ABC 60.

207 In Hänsgen's bibliography (*Antike Welten*, pp. 79–132), about 35 out of Kiepert's 549 maps are dedicated to Palestine, Sinai and Syria.

208 Kiepert to Ritter, Berlin, 21 December 1843, SBB-PK, Nachl. Carl Ritter, K 3, p. 33. The map, no. 39 in Hänsgen's list (*Antike Welten*, p. 83), exists also in the map department of Berlin State Library, Kart. D 6949.

209 Kiepert to Ritter, Berlin, 21 December 1843, SBB-SPK, Nachl. Carl Ritter, K 3, p. 33. Ritter, 'Neue Entdeckungen'; Robinson, 'Researches in Palestine'. Cf. Jones, 'British military surveys'; Goren, 'Pioneer surveying'; Goren, 'British surveyors'.

210 Samuel Wolcott, 'Maps of Palestine', *BSTR* 2/7 (August 1845); Joseph Hutchins Colton, *New Map of Palestine from the Latest Authorities; Chiefly from the Maps and Drawings of Robinson and Smith, with Corrections and Additions Furnished by the Rev. Dr. Robinson* (New York, 1845); Joseph Tracy, *Palestine and Mount Lebanon. Enlarged from Kiepert's Map, in Robinson and Smith's Biblical Researches, with Additions and Corrections, Principally from the Same Authorities* (Boston, [1845]); cf. Ritter, *Erdkunde von Asien*, pp. 87–8.

211 Letters from Kiepert to Ritter, SBB-SPK, Nachl. Carl Ritter, K 3, pp. 34–40.
212 Kiepert to Ritter, Weimar, 7 May 1849, SBB-SPK, Nachl. Carl Ritter, K 3, p. 40.
213 Nathan Michael Gelber, 'A pre-Zionist plan for colonizing Palestine. The proposal of a non-Jewish German-American in 1853' (trsl. Guido Kisch), *Historia Judaica*, 1/2 (1939); Goren, *Zieht hin*, pp. 262–9.
214 Charles Franz Zimpel, *Die Israeliten in Jerusalem. Eine Denkschrift gewidmet allen Israeliten* (Stuttgart, 1852); Charles Franz Zimpel, *Neue örtliche topographische Beleuchtung der heiligen Weltstadt Jerusalem, mit besonderer Rücksicht auf die Leidenstage unseres Herrn Jesu Christi und die Identität der heiligen Orte daselbst …* (Stuttgart, 1853).
215 Conrad Schick, 'Studien über Strassen und Eisenbahn-Anlagen zwischen Jaffa und Jerusalem, nebst Beschreibung der Gegend im Norden von Jaffa und der Ruinen von Mar Zacharias', *PGM* 13 (1867); Charles Franz Zimpel, *Straßen-Verbindung des Mittelländischen mit dem Todten Meere und Damascus über Jerusalem…* (Frankfurt am Main, 1865); Gelber, 'A Pre-Zionist Plan'; Goren, *Zieht hin*, pp. 266–9; Haim Goren and Bruno Schelhaas, 'Conrad Schick plant Straßen und Eisenbahnen', in P. Weigel et al. (eds), *Das Heilige Land in Gotha. Der Verlag Justus Perthes und die Palästinakartographie im 19. Jahrhundert* (Gotha, 2014). See in detail Chapter 3 in this volume.
216 Zögner, *Carl Ritter*, pp. 110–11. For Kiepert's further works of Palestine cartography: Goren, *Zieht hin*, pp. 102–4.
217 Ravenstein to Shaw, 12 May 1862, RGS-IBG Collections RGS Corr. Block 1861–70: Ravenstein, E.G.
218 Robinson to Smith, New York, 17 April 1841, ESP, ABC 60, Robinson, Edward.
219 Robinson, 'Researches in Palestine', p. 2.
220 The campaign received intensive studies, as Caesar E. Farah, *The Politics of Interventionism in Ottoman Lebanon, 1830–1861* (Oxford, London, New York, 2000); Muhammed H. Kutluoğlu, *The Egyptian Question (1831–1841). The Expansionist Policy of Mehmed Ali Paşa in Syria and Asia Minor and the Reaction of the Sublime Porte* (Istanbul, 1998). One of the many eyewitnesses' descriptions: August von Jochmus, *Der syrische Krieg und der Verfall des Osmanen-Reiches seit 1840. Aktenmässig dargestellt in officiellen, geheimen und vertraulichen Berichten und Urkunden …* (Frankfurt am Main, 1856).
221 Robinson to Smith, New York, 2 March 1846, ESP, ABC 60, Robinson, Edward.
222 Ibid.
223 Robinson to Smith, New York, 27 September 1848, ESP, ABC 60, Robinson, Edward.
224 William Francis Lynch, *Narrative of the United State's Expedition to the River Jordan and the Dead Sea* (Philadelphia, 1849); William Francis Lynch, *Official Report of the United States' Expedition to Explore the Dead Sea and the River Jordan* (Baltimore, 1852),citation vi. HLH, ABC 60, Lynch to Smith, Beirut, 14 and 29 July 1848, ESP, ABC 60: Lynch, W. F.; Robinson to Smith, New York, 29 November 1847, 27 September 1848, 24 January 1852, ESP, ABC 60: Robinson, Edward. Cf. Carl Ritter, *Der Jordan und die Beschiffung des Todten Meeres* (Berlin, 1850); Goren, *Dead Sea Level*, pp. 258–61; Jay Gomer Williams, *Times and Life*, pp. 310–1.
225 Robinson to Smith, New York, 31 March 1847, ESP, ABC 60, Robinson, Edward.
226 Robinson to Smith, London, 20 August and 8 September 1851, ESP, ABC 60, Robinson, Edward.
227 Robinson to Smith, New York, 19 November 1851, ESP, ABC 60, Robinson, Edward.
228 Robinson to Smith, New York, 19 November 1851; Berlin, 24 January 1852; Damascus, 3 June 1852, ESP, ABC 60, Robinson, Edward.
229 Robinson to Smith, Smyrna Harbour, 28 June 1852, ESP, ABC 60, Robinson, Edward.
230 Robinson to Ritter, Halle, 31 July 1851, UBF, Ritter NL 33/144/4; Robinson to Smith, Weimar, 5 September 1852, Halle, 2 October 1852, ESP, ABC 60, Robinson, Edward.
231 Robinson to Smith, Halle, 2 October 1852, ESP, ABC 60, Robinson, Edward. The volume (Ritter, *Erdkunde*, 16th part, Vol. VIII, 2nd unit) was published in 1855.
232 Tobler, *Bibliographia*, p. 166.
233 Josias Leslie Porter, *Map of Damascus, Hauran and the Lebanon Mountains* (London, 1855); Josias Leslie Porter, 'Memoir on the map of Damascus, Hauran, and the Lebanon mountains', *JRGS* 26 (1856); Josias Leslie Porter, *The Giant Cities of Bashan and Syria's Holy Places* (London, 1866); Josias Leslie Porter, *A Handbook for Travellers in Syria and Palestine; Including an Account of the Geography, History, Antiquities, and Inhabitants of these Countries …* (London, 1858). See Van de Velde's praising *Memoir*, pp. 18–21 and Aiken's brilliant study, *Scriptural Geography*, pp. 89–132.
234 Porter to Smith, Damascus, 21 June 1852 and 23 September 1852, ESP, ABC 60: Porter, Josias Leslie. See also Robinson to Smith, New York, 20 August 1853, ESP, ABC 60, Robinson, Edward.
235 Robinson to Smith, Halle, 2 October 1852, ESP, ABC 60, Robinson, Edward. For Symonds and his measurements cf. Jones, 'British military surveys'; Goren, *Dead Sea Level*, pp. 102–15, 231–47.

236 George Williams, *Holy City*, I, pp. vii–ix; Robinson, 'Topography of Jerusalem', BSTR, 3 (1846), pp. 414–15; Goren, *Zieht hin*, pp. 194–201; Haim Goren, ''Ernst Gustav Schultz referiert über Jerusalem', in P. Weigel et al. (eds), *Das Heilige Land in Gotha. Der Verlag Justus Perthes und die Palästinakartographie im 19. Jahrhundert* (Gotha, 2014).
237 Camille Callier, *Carte de la Syrie Méridionale et de la Palestine, dressée, en 1835, d'après les ordres du Directeur du Dépôt-général de la guerre Lieut.-Général Pelet* ([Paris], 1840); Lucie Bonato, 'Camille Callier. Un officier instruit de l'armée française qui explora Chypre en 1832', *Thetis* 10 (2003); Goren, *Dead Sea Level*, pp. 30–6 (including maps), 211–29.
238 Haim Goren, 'Lieutenant Symonds irrt bei Höhenmessungen', in P. Weigel et al. (eds), *Das Heilige Land in Gotha. Der Verlag Justus Perthes und die Palästinakartographie im 19. Jahrhundert* (Gotha, 2014).
239 Robinson to Smith, Halle, 2 October 1852, ESP, ABC 60, Robinson, Edward.
240 Van de Velde, *Memoir*; Van de Velde, *Plan of the Town and Environs of Jerusalem Constructed from the English Ordnance-Survey and Measurements of Dr. Titus Tobler, with Memoir by Dr. Titus Tobler* (Gotha and London, 1858); Charles William Meredith van de Velde, *Map of the Holy Land. From his own surveys in 1851 & 1852, from those made in 1841 by Majors Robe and Rochfort Scott, Lieut. Symonds ... and from the Results of the Researches made by Lynch, Robinson, Wilson, Burckhardt, Seetzen & c. Engraved by Eberhardt and by Stichardt* (Gotha, 1858). Cf. Haim Goren and Bruno Schelhaas, 'Colonial, religious and scientific mapping. The cartographic practice of Charles William Meredith van de Velde', in J. García-Álvarez and J. Carlos Garcia (eds), *History of Geography and Colonialism* (Lisbon, 2015); Sophie Perthus and Jutta Faehndrich, 'Visualizing the map-making process. Studying 19th century Holy Land cartography with MapAnalyst', *e-Perimetron* 8/2 (2013).
241 Robinson to Smith, Halle, 2 October 1852, ESP, ABC 60, Robinson, Edward. Van de Velde is mentioned in all letters from Robinson to Smith from 1852. See also van de Velde's letters from his excursion in Lebanon and his efforts to get Smith to check the spelling of his Arabic names: Van de Velde to Smith, Hasbeya, 10 January 1852; Damascus, 3 June 1852; Smyrna, 28 June 1852, ESP, ABC 60, General letters V W Y Z: C.W.M. van de Velde.
242 Kiepert, 'Topographische Beobachtung'
243 Murray to Robinson, Edinburgh, March 1856, ERP, AA1. This letter has been a bit damaged by water, so some of the words are impossible to decipher.
244 Engelmann, 'Physikalische Atlas'; Francis Herbert, 'The Royal Geographical Society's membership, the map trade, and geographical publishing in Britain 1830 to *ca* 1930. An introductory essay with listing of some 250 fellows in related professions', *Imago Mundi* 35 (1983), pp. 71, 82. See in detail Bruno Schelhaas's chapter in this book, Chapter 3.
245 Murray to Robinson, Edinburgh, March 1856, ERP, AA1.
246 Robinson to Smith, New York, 20 August 1853, ESP, ABC 60, Robinson, Edward.
247 Louis Felicien de Saulcy, *Narrative of a Journey Round the Dead Sea and in the Bible Lands in 1850 and 1851. Including an Account of the Discovery of the Sites of Sodom and Gomorrah*, I–II (London, 1854).
248 Paul-Adrien Bourdaloue, *Isthme de Suez. Parallèle entre le nivellement de 1799 at celui de 1847* (Bourges, 1856).
249 John Wilson, *The Lands of the Bible, Visited and Described in an Extensive Journey Undertaken with Special Reference to the Promotion of Biblical Research and the Advancement of the Cause of Philanthropy* (2 vols, Edinburgh and London, 1847); John Wilson, 'Map of the Holy Land with topographical plans of Jerusalem, Tyre, Sidon, Akka, and Beirut, constructed for Dr Wilson's "Lands of the Bible". By W. & A. K. Johnston, geographers and engravers to the Queen', in J. Wilson, *The Lands of the Bible, Vol. II* ; John Wilson, 'Map of the Peninsula of Mount Sinai and Arabia Petraea, constructed for Dr Wilson's "Lands of the Bible". By W. & A. K. Johnston, Geographers and Engravers to the Queen [insets: Valley and Rocks of Petra from Laborde; Central Group of the Mountains of Sinai]', in J. Wilson, *The Lands of the Bible, Vol. I.* Tobler (*Bibliographia*, p. 242) praised the lithography as '*recht schön*', but it has hardly any new information, as Wilson did not choose any new routes on his way.
250 Murray to Robinson, Edinburgh, March 1856, ERP, AA1.
251 Murray to Robinson, 7 July 1856, ERP, L. Correspondence to E.R. 1840–60.
252 RGS Secretary (signed for Norton Shaw) to Robinson, London, 19 November 1856, ERP, L. Correspondence to E.R. 1840–60.
253 Carl Helmuth, *Karte von Palästina nach Robinson, Smith und v. Schubert bearbeitet und in Stein gestochen von Carl Helmuth*, 1:600.000 (Halle, 1843); Carl Helmuth, *Plan von Jerusalem nach Robinson und Eli Smith entworfen und lithographiert von C. Helmuth* (Halle, 1843).
254 Tracy, *Palestine and Mount Lebanon.*

255 Karl Friedrich Robert Schneider, *Wandkarte zur biblischen Geschichte nach den neuesten Hülfsmitteln, besonders nach den Angaben und Karten von E. Robinson und E. Smith, K. v. Raumer, H. v. Schubert, H. Kiepert ..., bearbeitet. Gezeichnet von E. Braun u. E. Leeder, lithogr. von J. Wilhard in Dresden*, 1:500,000 (Dresden, undated).
256 Markus Kirchhoff, *Text zu Land. Palästina im wissenschaftlichen Diskurs 1865–1920* (Göttingen, 2005), pp. 336–46.
257 Rehav (Buni) Rubin, *Portraying the Land. Hebrew Maps of the Land of Israel from Rashi to the Early 20th Century* (Jerusalem, 2014) [Hebrew], pp. 227–71.
258 Tobler, *Bibliographia*, pp. 238–9.
259 Titus Tobler, *Planography of Jerusalem. Memoir to Accompany the New Ground-Plan of the City of Jerusalem ...* (Gotha, 1858), p. 13; Van de Velde, *Plan of the Town and Environs of Jerusalem.*
260 Ritter, *Comparative Geography*, II, p. 82; Ritter, *Erdkunde von Asien*, pp. 85–8.
261 Jay Gomer Williams, *Times and Life*, pp. 350–60.
262 No title on the map in the book.
263 The name 'Al Kuds' is written inside the real position of the city.
264 The maps in the British and the American editions are almost identical.
265 In this edition the biblical names are underlined and bolded and the name 'Al Kuds' is written outside its physical position.
266 The maps in the British and the American editions are almost identical.

CHAPTER 2

1 Edwin James Aiken, *Scriptural Geography. Portraying the Holy Land* (London, 2010); Haim Goren, *'Zieht hin und erforscht das Land'. Die deutsche Palästinaforschung im 19. Jahrhundert* (Göttingen, 2003)'; Markus Kirchhoff, *Text zu Land. Palästina im wissenschaftlichen Diskurs 1865–1920* (Göttingen, 2005), to name just a few.
2 See Haim Goren, Chapter 1 in this volume.
3 Kirchhoff, *Text zu Land*, p. 121.
4 Ibid, p. 132.
5 William M. Thomson, *The Land and the Book or Biblical Illustrations Drawn From the Manners and Customs, the Scenes and the Scenery of the Holy Land*. 2 vols (New York, 1859).
6 Kirchhoff, *Text zu Land*, p. 234.
7 Maurice Halbwachs, *Topographie légendaire des Évangiles en Terre Sainte. Étude de mémoire collective* (Paris, 1941).
8 Edward Said, *Orientalism* (London, 1978), while not without its flaws, remains as relevant for the Holy Land as for the entire Levant. For a more balanced study on the German tradition see Suzanne Marchand, *German Orientalism in the Age of Empire: Religion, Race, and Scholarship* (Cambridge, 2009).
9 Ulrich Johannes Schneider, 'Encyclopedias', in M. Middell (ed.), *Cultural Transfers, Encounters and Connections in the Global 18th Century* (Leipzig, 2014), p. 181.
10 Schelhaas and Wardenga, '"Inzwischen spricht die Karte für sich selbst"'. Transformation von Wissen im Prozess der Kartenproduktion', in S. Siegel and P. Weigel (eds), *Die Werkstatt des Kartographen. Materialien und Praktiken visueller Welterzeugung* (München, 2011).
11 Neil Asher Silberman, *Digging for God and Country. Exploration, Archaeology, and the Secret Struggle for the Holy Land, 1799–1917* (New York, 1982); Kirchhoff, *Text zu Land*, p. 144, pp. 196–229.
12 Kirchhoff, *Text zu Land*; Aiken, *Scriptural Geography*.
13 Sophie Perthus and Jutta Faehndrich, 'Visualizing the map-making process. Studying 19th century Holy Land cartography with MapAnalyst', *e-Perimetron* 8/2 (2013).
14 Meron Benvenisti, *Sacred Landscape. The Buried History of the Holy Land since 1948* (Berkeley, 2000), pp. 28–9. Later it became apparent that only a tenth of the collected names were antique, the rest was merely descriptive.
15 Haim Goren, 'Sacred, But Not Surveyed. 19th century surveys of Palestine', *Imago mundi* 54 (2002).
16 Haim Goren brought to my attention the fact that neither Seetzen nor Burckhardt had the Holy Land as their final destination but were both on their way to Africa. Neither did they lose their lives in Palestine but in Yemen (Seetzen, 1767–1811) and Egypt (Burckhardt, 1784–1817, died in Cairo).

17 Gudrun Krämer, *History of Palestine: From the Ottoman Conquest to the Founding of the State of Israel* (Princeton, 2011), pp. 71–100.

18 Yehoshua Ben-Arieh, *The Rediscovery of the Holy Land in the 19th Century* (Jerusalem, 1979, 2nd edn 1983).

19 Pierre Jacotin, *Carte topographique de l'Égypte et de plusieurs parties des pays limitrophes, levée pendant l'expédition de l'armée française* (47 sheets, Paris, 1810); Heinrich Berghaus, *Karte von Syrien, den Manen Jacotin's und Burckhardt's gewidmet* (Gotha, 1835). On Berghaus's role in Palestine cartography see Haim Goren, Chapter 1 in this volume and Plate 1.

20 My heartfelt gratitude to Petra Weigel and Sven Ballenthin of the Perthes Collection for their most generous help in locating and working with this correspondence and related archival materials. Sven Ballenthin even managed to find the earliest letters by Petermann pertaining to the van de Velde map in a previously not catalogued volume.

21 After about 1865 this changed to a more national perspective, a fact that must be kept in mind when studying the beginnings of Biblical geography and Palestine research.

22 Research on van de Velde has been sparse and far between. The most comprehensive studies are Dirk de Vries, 'Kaart van Palestina door C.W.M. van de Velde (1858): een autoriteitsstuk', in P. van der Krogt (ed.), *'Capita selecta' uit de geschiedenis van de kartografie* (Amersfoort, 1996), and Johannes H. Rombach, 'C.W.M. van de Velde (1818–1898), tekenaar reiziger enz.', *De negentiende eeuw. Documentatieblad Werkgroep 19e eeuw* 4/2–3 (1980). Haim Goren and Bruno Schelhaas, 'Colonial, religious and scientific mapping. The cartographic practice of Charles William Meredith van de Velde', in J. García-Álvarez and J. Carlos Garcia (eds), *History of Geography and Colonialism* (Lisbon, 2015), give an excellent overview of the latest state of research.

23 His birth entry shows that his first names were really Charles William Meredith and not the Dutch version 'Carel'; see the reproduction of his birth certificate (Fig. 2.1).

24 Marinus J.C. Klaassen, *Adelborstenopleiding te Delft–Medemblik–Breda, 1816–1857* (Den Helder, 1979), p. 79.

25 Philippus Christianus Molhuysen and P. J. Blok (eds), *Nieuw nederlandsch biografisch woordenboek* (10 vols, Leiden, 1911–37), pp. 1225–6.

26 Charles William Meredith van de Velde, *Map of the Holy Land. Constructed From His Own Surveys in 1851, 52 & 62 …* (Gotha, 1858, 1865).

27 Charles William Meredith van de Velde, *Memoir to Accompany the Map of the Holy Land* (Gotha, 1858), p. 1.

28 Kirchhoff, *Text zu Land*, p. 158, describes the 'military Bible Scholar' as characteristic of the later Palestine research after *c.*1865. On the PEF, see for example John James Moscrop, *Measuring Jerusalem. The Palestine Exploration Fund and British Interests in the Holy Land* (New York, 2000).

29 Charles William Meredith van de Velde, *Narrative of a Journey through Syria and Palestine in 1851 and 1852*, 2 vols (Edinburgh and London, 1854); Dutch: *Reis door Syrië en Palestina in 1851 en 1852* (Utrecht, 1854); German: *Reise durch Syrien und Palästina in den Jahren 1851 und 1852* (Leipzig, 1855–6).

30 James Strong and John McClintock, *The Cyclopedia of Biblical, Theological, and Ecclesiastical Literature* (New York, 1880) lists at least a dozen places whose identification dates back to van de Velde, as can be determined by a search of the electronic version, available at http://www.biblicalcyclopedia.com (accessed 30 June 2015).

31 De Vries, 'Kaart van Palestina', p. 16. In 1846 he had already published a widely acclaimed series of lithographies from his time on the island of Java: Charles William Meredith van de Velde, *Vues de Java, Sumatra, Borneo, Celebes et autres possessions hollandaises dans les Grandes Indes, dessinées d'après nature* (Amsterdam, 1846).

32 Charles William Meredith van de Velde, *Le pays d'Israël. Collection de cent vues prises d'après nature dans la Syrie et la Palestine* (Paris, 1857). For an analysis of his two versions of the Sea of Galilee, see Jutta Faehndrich, 'Charles William Meredith van de Velde zeichnet biblische Geschichte', in P. Weigel et al. (eds), *Das Heilige Land in Gotha. Der Verlag Justus Perthes und die Palästinakartographie im 19. Jahrhundert* (Gotha, 2014).

33 So Johannes H. Rombach, 'Two great figures in Red Cross history', *International Review of the Red Cross* 16 (1962); Rombach, 'C.W.M. van de Velde'; de Vries, 'Kaart van Palestina'; Guy Galazka, *A la découverte de la Palestine: voyageurs français en Terre sainte au XIXe siècle* (Paris, 2011). Rombach, 'Two great figures', explicitly mentions the diversity of van de Velde's œuvre, yet does not discuss its discrepancies.

34 Charles William Meredith van de Velde, 'C. W. M. van de Velde's letzte Reise in Palästina, 1861/62, und Bericht über die neue Auflage seiner Karte des Heiligen Landes', *Petermanns Geographische Mitteilungen* (*PGM*) 11 (1865).

35 Rombach, 'Two great figures'.

36 Klaassen, *Adelborstenopleiding*, p. 79; de Vries, 'Kaart van Palestina', p. 15.

37 Hommo Reenders found strong evidence of van de Velde's 'homosexual orientation'; Reenders, *Alternatieve zending: Ottho Gerhard Heldring (1804–1876) en de verbreiding van het christendom in Nederlands-Indië* (Kampen,

1991), p. 141, footnote 70. I was unable to consult the archival sources this is based on, but Reenders's study is otherwise sound and well-founded. As apparently van de Velde's sexual orientation had no influence on his cartographic work, it is mostly a biographical side note.

38 Charles William Meredith van de Velde, *Kaart van het Eiland Java te zamengesteld uit officiele bronnen door door C.W.M. van de Velde, Luitenant ter Zee* (Leiden, 1845).

39 De Vries, 'Kaart van Palestina', p. 16; Guillaume-Henri Dufour, *Topographische Karte der Schweiz, Carte topographique de la Suisse, Carta topografica della Svizzera* (Geneva, 1845–65). Both men knew each other from Geneva's Protestant circles; like van de Velde, Dufour had a military background and later was among the founders of the Red Cross.

40 The highest resolution version is available on the Amsterdam Free University library website at http://imagebase.ubvu.vu.nl/getobj.php?ppn=330046055 (accessed 2 June 2015).

41 'Obituary', *The Times*, 24 March 1898, p. 10.

42 Van de Velde, *Narrative of a Journey through Syria and Palestine*, I, p. 164 (first meeting) and *passim*. The boy only goes by the first name 'Philip' and was a pupil of the American missionary school.

43 Van de Velde, *Memoir*, pp. 3–4.

44 Ibid., p. 23.

45 Titus Tobler, *Bibliographia geographica Palaestina. Zunächst kritische Übersicht gedruckter und ungedruckter Beschreibungen der Reisen ins heilige Land* (Leipzig, 1867), p. 243: 'decidedly the best that we now know' (1867).

46 Joachim Perthes, 'Fünf Generationen', in *Justus Perthes 1785–1935, Hauptkatalog 1935* (Gotha 1935), p. XV.

47 Van de Velde to Petermann, Utrecht 8 December 1854, FBG SPA ARCH MFV 9, f. 351–2. In the following, the van de Velde correspondence (FBG SPA ARCH MFV 9) will only be referred to by folio numbers. Petermann and van de Velde corresponded in English. To the publishers van de Velde wrote in French, unless otherwise specified; they answered in German. All English quotes from the correspondence with Bernhardt Perthes and his successors are my own translations.

48 The full title until Petermann's death was *Mittheilungen aus Justus Perthes Geographischer Anstalt über wichtige neue Erforschungen auf dem Gesamtgebiet der Geographie von Dr. A. Petermann*, later also known as *PGM* (*Petermanns Geographische Mitteilungen*).

49 Van de Velde to Petermann, Edinburgh, 23 May 1855, f. 355–6.

50 Ibid.

51 Yolande Jones, 'British military surveys of Palestine and Syria 1840–1841', *The Cartographic Journal* 10/1 (1973); Haim Goren 'Sacred, But Not Surveyed' and 'Lieutenant Symonds irrt bei Höhenmessungen', in P. Weigel et al. (eds), *Das Heilige Land in Gotha. Der Verlag Justus Perthes und die Palästinakartographie im 19. Jahrhundert* (Gotha, 2014). See also Chapter 3 in this volume.

52 Jacotin, *Carte topographique de l'Égypte*.

53 Anne Godlewska, *The Napoleonic Survey of Egypt. A Masterpiece of Cartographic Compilation and Early Nineteenth-Century Fieldwork* (Toronto, 1988).

54 Initially a 5½-inch theodolite, later a superior 7-inch one; Jones, 'British Military Surveys', pp. 31–5; the triangulations are reconstructed on p. 35. A sketch of the triangulation network was later communicated to August Petermann by Admiral Washington, hydrographer of the Royal Navy; reproduced in Goren, 'Lieutenant Symonds', p. 70; see also Bruno Schelhaas, Chapter 3 in this volume.

55 Adolf Stieler, *Hand-Atlas über alle Theile der Erde* (Gotha, 1817–1945), a prime product of the Perthes publishing house.

56 Petermann to Perthes, London, 7 June 1855, FBG SPA Arch PGM 540, ff. 1–2. Haik Thomas Porada deserves full credit for helping me decipher critical words in Petermann's tricky handwriting.

57 'Il est inutile de vous démonstrer l'importance de ma carte.' Letter by van de Velde to Justus Perthes publishers (in the following referred to in short as 'Perthes'), London, 6 June 1855, f. 1.

58 On Tobler see Goren, *Zieht hin*, pp. 222–43.

59 Van de Velde to Perthes, Horn bei Rorschach (Tobler's house), 22 August 1855, ff. 12–13, here 12r. The Jerusalem map based on the Royal Engineers' work was published in Williams, George, *The Holy City. Historical, Topographical, and Antiquarian Notices on Jerusalem*, I–II (London, 1849). It has two cartographic errors that make it easy to follow the map's genealogy: the wall of the Haram is drawn with protruding corners, which were really private houses built along the wall, and the position of the entire Haram and eastern city wall is aligned to north instead of slightly north-north-west. Henry Lange used it for Carl Ritter's *Erdkunde* (Henry Lange, *Skizze eines Grundrisses von Jerusalem. Zu C. Ritter's Allg. Erdkunde. Theil XVI, Palaestina* (Berlin, 1852), pp. 297–508, and van de Velde in *Narrative* I. Eventually the Jerusalem map by Tobler and van de Velde bore the same title as the Royal Engineers' map (*Plan of Town and Environs of Jerusalem*). See also

Jones, 'British Military Surveys'; Goren, 'Sacred, But Not Surveyed'. For the Jerusalem map see Haim Goren and Petra Weigel's 'Heinrich Lange kartiert Carl Ritters Forschungen zu Jerusalem' and Jutta Faehndrich's 'Charles William Meredith van de Velde entwirf mehr als ein Kartenbild' and 'Titus Tobler sammelt Straßennamen' all three in P. Weigel et al. (eds), *Das Heilige Land in Gotha. Der Verlag Justus Perthes und die Palästinakartographie im 19. Jahrhundert* (Gotha, 2014).

60 Titus Tobler, *Planography of Jerusalem. Memoir to Accompany the New Ground-Plan of the City of Jerusalem ...* (Gotha, 1858). FBG SPA ARCH MFV 141 (Titus Tobler correspondence).

61 Van de Velde to Perthes, Clarens, 21 September 1855, FBG SPA ARCH MFV 9, ff. 14–5, here f. 14r. There are just as many pointed remarks about van de Velde in the Tobler correspondence, yet overall their letters show how much they respected each other. Unfortunately we cannot elaborate on their relationship here.

62 William Thomas Gidney, *The History of the London Society for Promoting Christianity Amongst the Jews. From 1809 to 1908* (London, 1908), p. 213: 'In 1849, Captain Henry L. Layard was appointed Lay Secretary.' According to Layard's obituary (see pp. 337–8), he died on 24 February 1871. He had been Captain in the 97th Regiment, serving in India until his retirement in 1835, lived in Ceylon in 1841–9, and is said to have been a 'great friend of the Jewish people' who often visited the Holy Land and the missions. He was most likely a cousin of the Niniveh explorer Sir Austen Henry Layard, their common grandfather being Rev. Charles Peter Layard (1748–1803).

63 Josias Leslie Porter, *Five Years in Damascus* (2 vols, London, 1855), wrongly attributed to John Leech Porter by Reinhold Röhricht, *Bibliotheca Geographica Palaestinae. Chronologisches Verzeichnis der auf die Geographie des Heiligen Landes bezüglichen Literatur von 333 bis 1878 und Versuch einer Cartographie* (Berlin, 1890, repr. Jerusalem, 1963, ed. David H.K. Amiran). On Porter see Aiken, *Scriptural Geography.*

64 Van de Velde to Perthes, Clarens, 21 September 1855, ff. 14–15, here 15v.

65 Van de Velde to Perthes, Geneva, 13 November 1855, ff. 20–2, here 20v.

66 Valentina Anker, 'Calame, Alexandre', in *Historisches Lexikon der Schweiz*, available at http://www.hls-dhs-dss.ch/textes/d/D21996.php (accessed 23 June 2015).

67 Later the author of Thomson, *The Land and the Book.*

68 Van de Velde to Perthes, Geneva, 13 November 1855, ff. 20–2.

69 Namely Arthur Penrhyn Stanley, *Sinai and Palestine, in Connection With Their History* (London, 1856). Available online at http://www.biblicalcyclopedia.com.

70 Van de Velde to Perthes, Geneva, 29 December 1855, ff. 26–7.

71 Van de Velde to Perthes, Geneva, 14 January 1856, ff. 41–2, here 41v.: 'de venir m'etablir à Gotha et m'engager à votre belle et importante entreprise'.

72 Ibid., f. 42.

73 Van de Velde to Perthes, Geneva, 6 February 1856, ff. 36–8.

74 Van de Velde to Perthes, Geneva, 26 February 1856, f. 33–4.

75 Draft letter by Perthes to van de Velde, 2 March 1856, ff. 39–40. Special thanks to Teresa Kilian for helping me read this scribble.

76 Correpondence with both has survived in the Perthes Collection (FBG SPA ARCH MFV 134 Stichart, Heinrich; FBG SPA ARCH MFV 051 Eberhardt, Hermann), but does not contain information on the issue.

77 Ibid., 39r. From a later letter, we know that the engraving was done on two plates, one depicting lines, the other the relief ('Terrain'); van de Velde to Perthes, Basle, 27 October 1862, ff. 307–8, asking for a print 'without mountains' for the second edition, f. 307v. So only the line engraver responded to the reproach.

78 Ibid.

79 Van de Velde to Perthes, Geneva, 26 February 1856, ff. 33–4.

80 Van de Velde to Perthes, Geneva, 31 March 1856, ff. 45–8, here f. 45r.

81 Ibid., ff. 45r–6r.

82 Ibid., f. 46r.

83 Ibid., f. 47r.

84 Van de Velde, *Memoir*, p. 66. According to the World Geodetic System 1984 (WGS 84), Gaza City today is located at 31°31'N and 34°27'E. I have to thank Eric Losang for trying to help me solve this mystery, though to no avail.

85 Van de Velde to Perthes, Geneva, 31 March 1856, ff. 45–8, here f. 47.

86 Ibid., f. 49v.

87 Ibid., f. 48. On Kiepert's maps for Robinson and Smith, see Haim Goren, Chapter 1 in this volume. It seems that van de Velde indeed informed Porter or his publisher; the Edward Robinson Papers hold a letter by John Murray to Robinson of 7 July 1858: 'I hear to my regret that Porters Map of the Lebanon – is far out from his

having computed by the Magnetical, not the real North!! Do not let your Map Maker fall into P[orter]s mistakes – Van de Veldes map will not be out till the end of this year.' (HBA ERP L1 London Correspondence 1840–58, no folio number, spelling as in the original, transcription by Bruno Schelhaas).

88 Van de Velde to Perthes, London, 12 July 1856, ff. 51–2.

89 Van de Velde to Perthes, Haarlem, 19 August 1856 (ff. 55–6) and Utrecht, 4 September 1856 (ff. 53–4). A research project on van de Velde's French famous contemporary Honoré Daumier (1808–79), caricaturist and lithographer, lists prices in mid-nineteenth-century France: for 250 francs, one could buy a one-week trip from Paris to London, all inclusive, or pay the yearly rent for Daumier's house in Paris. A good piano cost 450 francs, a winter coat 50 francs, a full menu at a luxury restaurant 25 francs; Lilian and Dieter Noack, 'Cost of Living in Daumier's Time', available at http://www.daumier.org/176.0.html (accessed 25 June 2015).

90 Van de Velde to Perthes, Haarlem, 25 September 1856, ff. 57–9.

91 Van de Velde to Perthes, Paris, 17 November 1856, ff. 59–60.

92 Edward Robinson, Eli Smith et al., *Later Biblical Researches in Palestine and the Adjacent Regions. A Journal of Travels in the Year 1852* (London, 1856); see Haim Goren, Chapter 1 in this volume.

93 He met them first in Jerusalem in April 1852 (van de Velde, *Narrative* II, pp. 248–9), then again in Nablus in May, when they made a joint excursion to Pella east of the Jordan (p. 329).

94 Van de Velde to Perthes, Paris, 30 December 1856, f. 63–4, also mentioned in van de Velde, *Memoir*, pp. 5–6; cf. Charles Rochfort Scott, *Map of Syria, Constructed From the Surveys and Sketches of the Undermentioned Officers in that Country in 1840 and 1841, by Major C. Rochfort Scott, R. Staff-Corps, Under Whose General Direction the Work Was Undertaken, Major F. H. Robe, 87th Fusileers and R. Wilbraham, 7th Fusileers, and Lieut. J. F. A. Symonds R[oyal]. Engineers* (London, 1846). On the importance of this map see Goren, 'Sacred, But Not Surveyed', p. 90.

95 Stanley, *Sinai and Palestine*; Porter, *Five Years*.

96 Van de Velde to Perthes, Paris, 1 December, ff. 61–2, and 30 December 1856, ff. 63–4.

97 Heinrich Kiepert, *Übersichts-Karte von Palästina und Phönicien, vorzüglich nach den Messungen und Beobachtungen von Ed. Robinson und Eli Smith, und mit Benutzung aller bis jetzt erschienen Reisewerke neu construirt von H. Kiepert zur Erläuterung von Prof. Ed. Robinson's zweiter Reise in Palästina* (Berlin, 1857). The map to accompany the first edition had a scale of 1:400,000; the sketch map by Kiepert's hand in the Staatsbibliothek zu Berlin even 1:200,000; see Haim Goren's map bibliography in Chapter 1 of this volume.

98 Van de Velde to Perthes, Paris, 3 March 1857, ff. 65–7, here 67r.

99 Van de Velde to Perthes, Paris, 30 March 1857, ff. 68–9, here, f. 68r.

100 Ibid., f. 69r.

101 Van de Velde to Perthes, Paris, 4 April 1857 (f. 72) and Utrecht, 29 May 1857 (ff. 73–4).

102 Van de Velde to Perthes, Paris, 17 October 1857, ff. 75–6.

103 Ibid.

104 Perthes, 'Fünf Generationen'.

105 This was general practice, as letters by the English publishing houses Williams & Norgate and Alexander Keith Johnston and the German publisher Breitkopf & Härtel in the same correspondence show – all signed only with company name.

106 The Perthes family, with its origins in Hamburg, was staunchly Protestant.

107 Van de Velde to Perthes, Paris, 9 November 1857, ff. 77–8.

108 Ibid., margin of f. 77r. A German version of the memoir was never published.

109 Van de Velde to Müller, Paris, 8 December 1857, f. 79.

110 Van de Velde to Perthes, Paris, 23 January 1858, f. 80, the last word underlined in the original.

111 Ibid.

112 Van de Velde to Perthes, Paris, 17 March 1858, ff. 102–3.

113 On map memoirs and commentaries see Alexander Schunka, 'Das Rohe, das Gekochte – und das Kochrezept. Kartenkommentare des 19. Jahrhunderts als historische Quellen', in S. Siegel and P. Weigel (eds), *Die Werkstatt des Kartographen. Materialien und Praktiken visueller Welterzeugung* (München, 2011).

114 Van de Velde to Perthes, Paris, 26 April 1858, ff. 111–14.

115 Van de Velde, *Narrative* II, pp. 486–9 (entitled 'A Last Trial').

116 Van de Velde to Perthes, Utrecht, 29 May 1857 (ff. 73–4), to name only one.

117 Tobler to Perthes, 11 April 1857, FBG SPA MFV 141, f. 27 ('wenn er nur seine kräfte nicht zu sehr zersplittert'); undated letter, *c.*December 1858, f. 71 ('sehr schwächlicher konstitution, mit allerlei leiden beinahe immer behaftet,'); 11 December 1862, f. 113 ('Meine freude ist groß, dass Lieut. van de Velde zum zweiten mal nach Palästina hinpilgern will, überzeugt, dass er manches frische, für die wissenschaft von werth, bringen wird.

Möchte nur sein guter wille durch eine kräftigere konstituzion, deren mangel ihn zu manchen, einen kräftigen mann anwidernden[!] sentimentalien[!!] verleitet, unterstützt werden.').
118 Van de Velde to Perthes, Paris, 18 May 1858, ff. 115–16.
119 Van de Velde to Perthes, Grion, 12 July 1858, ff. 131–2.
120 Perthes (Müller) to van de Velde, 16 July 1858, f. 133.
121 Stieler, *Hand-Atlas.*
122 Perthes (Müller) to van de Velde, Grion, 19 July 1858, ff. 137–8.
123 Breitkopf & Härtel to Perthes, Leipzig, 5 August 1858, f. 147.
124 Van de Velde to Perthes, Champéry, 24 August 1858, ff. 150–1.
125 Van de Velde to Müller, Glyon (near Montreux), 21 September 1858, ff. 156–7.
126 Ibid., f. 157r.
127 Perthes to van de Velde, 18 November 1858, ff. 173–4.
128 Williams & Norgate to Perthes, London, 8 October 1858, f. 165.
129 Perthes to van de Velde, 18 November 1858, ff. 173–4.
130 Van de Velde to Perthes, Nice, 13 November 1858, ff. 171–2.
131 Perthes to van de Velde, 19 November 1858, ff. 175–6.
132 Berghaus was supposed to make the maps for Robinson and Smith before Heinrich Kiepert took over; see Haim Goren, Chapter 1 in this volume.
133 Van de Velde to Perthes, Nice, 26 November 1858, ff. 180–1.
134 Perthes (Müller) to van de Velde, 10 January 1859, ff. 182–4.
135 Van de Velde to Perthes, Nice, 17 January 1859, ff. 188–9.
136 On Fallmerayer and von Raumer as Palestine explorers see Goren, *Zieht hin*, pp. 139–47.
137 Perthes (Müller) to van de Velde, Nice, 28 January 1859, ff. 192–4.
138 Emmanuel Guillaume Rey, 'Examen de quelques parties de la carte de la Palestine de M. van de Velde [review].' *Bulletin de la Société de géographie de Paris* 17 (1859).
139 Perthes (Müller) to van de Velde, 23 February 1859, ff. 202–3, here f. 203: 'Unter den vielen großen Unternehmungen [on the geography of Palestine] etc. nimmt der Altas des gelobten Landes von Van de Velde einen der wichtigsten Plätze ein', quoting A. von Humboldt.
140 Heinrich Berghaus, *Karte von Syrien* and *Geographisches Memoir zur Erklärung und Erläuterung der Karte von Syrien* (Gotha, 1835) (see Plate 1); Carl Zimmermann, *Atlas von Palaestina und der Sinai-Halbinsel; zu C. Ritters Erdkunde Band XIV-XVI* (Berlin, 1850); Kiepert did not publish a map of the region in 1852, see the bibliography by Haim Goren, Chapter 1 in this volume.
141 Anon., 'Geography and Biography of the Old Testament', *The London Quarterly Review* (American edn) 106 (October 1859).
142 Heinrich Ewald, 'Übersicht der 1859–60 erschienenen Schriften', *Jahrbücher der Biblischen Wissenschaft* 10 (1860), pp. 264–5: 'Dieses Werk ist nun so großartig angelegt und auch künstlerisch so schön ausgeführt daß es alle früheren ähnlichen weit übertrifft.'
143 Friedrich August Arnold, 'Map of the Holy Land [review]', *Neue Jahrbücher für Philologie und Pädagogik* 80 (1859), p. 256.
144 Victor Adolphe Malte-Brun, 'Map of the Holy Land [review]', *Nouvelles annales des voyages* 18/6 (1859), pp. 353–5.
145 The War Office to Justus Perthes, Pall Mall, 22 January 1859, ff. 190–1.
146 Van de Velde to Perthes, Lausanne, 14 June 1859, ff. 200–1.
147 Perthes to van de Velde, 20 June 1859, ff. 222–5.
148 Christian Karl Josias von Bunsen (1791–1860) was a philologist, theologian, Prussian diplomat and a remarkable orientalist networker; see Goren, *Zieht hin*, pp. 105–6.
149 Gustav Parthey, heir to the Nicolaische Buchhandlung; see Haim Goren, Chapter 1 in this volume.
150 Van de Velde to Müller, Petit-Saconnex, 9 July 1859, ff. 226–7.
151 Perthes to van de Velde, 21 December 1859, ff. 235–6.
152 Van de Velde to Perthes, Lausanne, 30 December 1859, ff. 233–4, here 233v.
153 Ibid.
154 Perthes (Müller) to van de Velde, 11 January 1860, f. 237.
155 Ibid. and Titus Tobler, *Titus Toblers dritte Wanderung nach Palästina im Jahre 1857. Ritt durch Philistäa, Fußreisen im Gebirge Judäas und Nachlese in Jerusalem* (Gotha, 1859).
156 Van de Velde to Müller, Lausanne, 17 January 1860, ff. 241–3.
157 Müller to van de Velde, 23 January 1860, f. 243.

158 Van de Velde to Perthes, Lausanne, 28 January 1860, ff. 248–9.
159 Van de Velde to Perthes, Lausanne, 13 February 1860, ff. 252–3.
160 Van de Velde to Perthes, Lausanne, 21 February 1860, ff. 255–6: 'you have tried to remunerate me as generously as possible and I would be glad if our relations continued for a long time.'
161 Van de Velde to Perthes, Lausanne, 6 May 1860, ff. 259–60; Perthes to van de Velde, 19 June 1860, ff. 267–8.
162 A. K. Johnston to Perthes, Edinburgh, *c.*July 1860, ff. 264–5 (English in the original, spelling not corrected); Perthes (Müller) to Johnston, Gotha, 20 July 1860, f. 266.
163 Perthes (Müller) to van de Velde, 19 June 1860, ff. 267–8.
164 Van de Velde to Perthes, Champéry, 20 July 1860, ff. 269–70; Perthes to van de Velde, 2 December 1860, ff. 272–3.
165 Van de Velde to Perthes, Champéry, 20 July 1860, ff. 269–70.
166 Van de Velde to Perthes, Ormonts-Dessus, 25 June 1861, ff. 280–1.
167 On the Syrian Asylums Committee see Moscrop, *Measuring Jerusalem*, pp. 48–51. One of its founding members was Sir Culling Eardsley, a subscriber to the *Map of the Holy Land*, ibid., p. 50.
168 Van de Velde to Perthes, Baden-Baden, 30 October 1861, ff. 283–4.
169 Van de Velde to Perthes, Paris, 11 October 1861, f. 286, Haarlem, 21 November 1861, ff. 289–90, Marseilles, 30 November 1861, ff. 294–5.
170 On the Spittler Mission in Jerusalem and its somewhat controversial shop see Alex Carmel, 'C. F. Spittler and the activities of the pilgrims mission in Jerusalem', in Gad G. Gilbar (ed.), *Ottoman Palestine 1880–1914: Studies in Oconomic and Social History* (Leiden, 1990).
171 Van de Velde to Perthes, Jerusalem, 6 February 1862, ff. 291–2.
172 Van de Velde, *Memoir*, p. 25–6: 'Here and there the words *not examined* will be found on the Map, indicating that we had no opportunity of either visiting such places or obtaining sufficiently correct information about them from others.'
173 Van de Velde, 'Van de Velde's letzte Reise in Palästina, 1861/62'.
174 Van de Velde, *Narrative* I, pp. 410–16.
175 For an in-depth analysis, including a comparison of location accuracy of both editions, see Perthus and Faehndrich, 'Visualizing the map-making process'.
176 Van de Velde to Perthes, London, 23 May 1862, ff. 296–7, here 296r, spelling as in the original. On Sandreczki see Haim Goren, Chapter 1 in this volume, pp. 5–54. The maps mentioned are Lucien A. Blondel, *Carte du Liban d'après les reconnaissances de la brigade topographique du corps expéditionnaire de Syrie en 1860–1861, dressée au Depôt de la Guerre, etant directeur le General Blondel, sous le ministère de S. E. le Marechal Comte Randon* (Paris, 1862). Mansell, Arthur L. and Thomas A. Hull (1864). *Mediterranean. Coast of Syria, Sheet 2 and 3* (London, 1864). On Johann Gottfried Wetzstein, Prussian consul at Damascus 1848–62, and Georg Rosen, Prussian consul at Jerusalem, see Goren, *Zieht hin*, pp. 185–93. The map to Wetzstein's travels was made by Kiepert, 'Note über die Construction', pl. IV; see Haim Goren's map bibliography in Chapter 1 of this volume.
177 Van de Velde to Perthes, London, 23 May 1862.
178 Perthes (Müller) to van de Velde, 21 June 1862, ff. 298–9.
179 Van de Velde to Perthes, Geneva, 16 July 1862, ff. 300–1.
180 Perthes (Müller) to van de Velde, 26 July 1862, ff. 303–4.
181 Van de Velde to Perthes, Basle, 27 October 1862, ff. 307–8.
182 The list of materials supplied for the second edition is reproduced as Fig. 2.6. Perthes (Müller) to van de Velde, 11 November 1862, f. 311.
183 Van de Velde to Perthes, Haarlem, 2 December 1862, ff. 314–15.
184 On the survey see Goren, 'Sacred, But Not Surveyed', p. 95; for a reproduction of one of the maps, including an in-depth analysis, see Sven Ballenthin, 'Commander Mansell vermisst Küsten und Meerestiefen', in P. Weigel et al. (eds), *Das Heilige Land in Gotha. Der Verlag Justus Perthes und die Palästinakartographie im 19. Jahrhundert* (Gotha, 2014).
185 Van de Velde, 'Reise in Palästina 1861/62', p. 297. See Goren, *Zieht hin*, pp. 188–90.
186 Van de Velde to Perthes, Haarlem, 2 December 1862, ff. 314–15.
187 Perthes (Müller) to van de Velde, 9 December 1862, ff. 316–17.
188 Van de Velde to Perthes, Haarlem, 14 December 1862, ff. 318–19.
189 Van de Velde to Perthes, Haarlem, 29 April 1863, ff. 320–1: 'qu'il verrait de parcourir la Palestine avec quelques de ses officiers pour collationner ma carte; qu'il avait trouvé ma carte d'une exactitude merveilleuse, et qu'il en était tellement satisfait, qu'il m'offrait toutes ses propres observations astronomiques et triangulairent.' Washington had already stated that 'Van de Velde's map of Palestine [...] was found to be

generally correct, and the best map of the country published.' (John Washington, 'Admiralty Surveys', *Journal of the Royal Geographical Society* 31 (1861), p. cxxxix).

190 Van de Velde to Perthes, Haarlem, 29 April 1863, ff. 320–1, here 320v: Mansell's results 'me permettent d'établir ma carte corrigée d'une manière durable, sans qu'il soit nécessaire de la retoucher plus jamais'.

191 Perthes (Müller) to van de Velde, 2 June 1863, ff. 322–3.

192 Blondel, *Carte du Liban*.

193 Van de Velde to Perthes, Geneva, 9 December 1863, ff. 326–7.

194 Perthes (Müller) to van de Velde, 30 December 1863, f. 328.

195 Van de Velde to Perthes, not dated, *c.*March 1864, ff. 339–40.

196 Van de Velde to Perthes, 23 March 1864, ff. 343–4.

197 Van de Velde to Müller, Geneva, 23 March 1864, f. 345.

198 Van de Velde to Müller, Paris, 2 April 1864, f. 347; see Jutta Faehndrich, 'Charles William Meredith van de Velde entwirf mehr als ein Kartenbild', in P. Weigel et al. (eds), *Das Heilige Land in Gotha. Der Verlag Justus Perthes und die Palästinakartographie im 19. Jahrhundert* (Gotha, 2014).

199 Most likely Grand Duke Konstantin Nikolayevich of Russia (1827–92), second son of Tsar Nicholas I and himself a navy admiral as well as enthusiastic geographer.

200 Van de Velde to Perthes, Paris, 12 November 1864, ff. 362–5.

201 Charles William Meredith van de Velde, *Carte de la Terre Sainte pour servir à l'étude des Saintes Ecritures* (Paris, 1865).

202 Van de Velde to Perthes, Paris, 5 October 1864, ff. 354–5.

203 Perthes (Müller) to van de Velde, 12 October 1864, ff. 356–7.

204 Perthes (Müller) to van de Velde, 1 May 1865, ff. 385–6.

205 Ibid.

206 Perthes to van de Velde, 8 May 1865, f. 389.

207 Van de Velde to Perthes, Paris, 2 February 1865, ff. 369–70.

208 Moscrop, *Measuring Jerusalem*, pp. 53–8; Haim Goren, 'Scientific organizations as agents of change. The Palestine Exploration Fund, the Deutsche Verein zur Erforschung Palästinas and nineteenth-century Palestine', *Journal of Historical Geography* 27 (2001).

209 Van de Velde to Perthes, Paris, 16 February 1865, ff. 371–2; excerpts of undated letter by RGS ibid., f. 371v. Sir Henry James (1803–77) was director-general of the Ordnance Survey 1854–75. Eventually the Survey of Western Palestine commenced in 1871 and was carried out by the PEF (founded in May 1865); see Moscrop, *Measuring Jerusalem*, p. 67. It is uncertain if there was an early financial contribution by the RGS. On a final note, van de Velde's map was not officially endorsed by the RGS.

210 Perthes (Müller) to van de Velde, 12 May 1865, f. 392.

211 Van de Velde to Perthes, Glyon, 10 May 1865, ff. 390–1; Geneva, 17 June 1865, ff. 396–7; Interlaken, 7 July 1865, ff. 398–9; Haarlem, 11 November 1865, ff. 409–10.

212 Perthes (Müller) to van de Velde, 1 August 1865, ff. 400–2.

213 Van de Velde to Perthes, Haarlem, 11 November 1865, ff. 409–10. Rosen's letter did not survive as van de Velde asked for it to be sent back to him (f. 410v).

214 Perthes to van de Velde, 20 November 1865, ff. 411–12.

215 Van de Velde to Perthes, Geneva, 18 April 1866, f. 422.

216 Sketch calculation of Bernhardt Perthes's hand, not dated, *c.*1855, FBG SPA ARCH MFV 9, ff. 6–7.

217 FBG SPA ARCH FFA VAB 1 (1854–69) and 2 (1869–74). Sven Ballenthin was so kind to point these books out to me.

218 Tobler to Perthes, Horn, 23 December 1865, FBG SPA ARCH MFV 141, f. 120.

219 Petermann to Arnold, 28 November 1866, FBG SPA ARCH *PGM* Konvolut As 29, f. 43f. On Petermann's Palestine cartography see Bruno Schelhaas, Chapter 3 in this volume.

220 Van de Velde to Perthes, Geneva, 14 January 1856, f. 41f.

221 Perthes to van de Velde, 19 July 1858, f. 137f.

222 Perthes to van de Velde, 19 November 1858, f. 175f.

223 Jutta Faehndrich, 'Theodor Menke konzentriert Bibelkarten', in P. Weigel et al. (eds), *Das Heilige Land in Gotha. Der Verlag Justus Perthes und die Palästinakartographie im 19. Jahrhundert* (Gotha, 2014).

224 Felix Driver, *Geography Militant. Cultures of Exploration and Empire* (Oxford, 2001), p. 12.

225 Tobler to Perthes, not dated, *c.*August 1857, FBG SPA ARCH MFV 141, f. 30: 'hr. van de Velde schlägt in den büchern nicht gerne nach', Tobler to Perthes, not dated, *c.*January 1858, f. 71: 'van de Velde [ist] zu abfassung des onomasticum oder der biblischen geographie nicht geeignet, allervörderst weil ihm die kenntniss der

alten sprachen abgeht.' Spelling as in the original. Tobler was actually both, a learned erudite and an expert field researcher who traversed Palestine on foot many times.

226 Jones, 'British Military Surveys', p. 29.

227 See Perthus and Faehndrich, 'Visualizing the map-making process', p. 63.

228 Driver, *Geography Militant*, p. 35–6: 'Yet [...] the construction of a reliable topographical map was as likely to require the compilation of maps in the office as it was to rely on an accurate instrumental survey in the field, even after the adoption of trigonometrical methods.' See also Matthew H. Edney, *Mapping an Empire. The Geographical Construction of British India, 1765–1843* (Chicago, 1997).

229 Haim Goren, *Dead Sea Level. Science, Exploration and Imperial Interests in the Near East* (London, 2011), pp. 102–5 and 231–7.

230 Jones, 'British Military Surveys', p. 38: 'It is difficult to make an independent assessment of the accuracy of Symonds' work, not only because it is not available today, but also because it was incorrectly used in the printed map.' This is based on van de Velde's comparison of the original notes he consulted at the Admiralty with the finished map; van de Velde, *Memoir*, p. 6.

231 Edward Robinson, 'Depression of the Dead Sea and of the Jordan Valley', *JRGS* 18 (1848); see van de Velde, *Memoir*, p. 168.

232 See Petermann's verdict in a letter to Ritter of June 1849, quoted by Bruno Schelhaas on p. 117.

233 Perthus and Faehndrich, 'Visualizing the map-making process', pp. 76–9.

234 Ben-Arieh, *Rediscovery*, p. 147.

235 Robinson to Smith, Smyrna Harbour, 28 June 1852, HLH ABC 60, ff. 3,809–11, here 3,810: 'Mr Van de Velde is talking of taking a copy of your former notes, & states he had before. I have replied to him to write to you; please give me explicit directions. He is beginning to assume a little more of a dictatorial tone.' Robinson to Smith, Vienna, 9 August 1852, HLH ABC 60, ff. 3,812–15, here 3,814: 'The view you take in that letter is just what has lain upon my own mind all the time. But as in all the conversations I had with Mr v. de V. he repeatedly affirmed that you had given him permission to copy and use your notes, I could only say to him that in such case I could not object.' Robinson to Smith, Weimar, 5 September 1852, ff. 3,816–19, here 3,816: 'Every day convinces me more & more that Mr Van de Velde could have made little or no use of your journals.' All transcripts by Bruno Schelhaas. Cf. van de Velde, *Narrative* II, pp. 329–67, about the excursion to Pella; p. 249 for his thoughts on a potential cooperation.

236 Van de Velde, *Memoir*, p. 116. See Perthus and Faehndrich, 'Visualizing the map-making process', p. 71.

237 Eberhardt to Perthes, Eisenach, 14 September 1855, FBG ARCH MFV 51, f. 170.

238 This very liberal use of material was not uncommon at that time; see Richard J. Scully, '"North Sea or German Ocean?" The Anglo-German cartographic freemasonry, 1842–1914', *Imago Mundi* 62/1 (2010).

239 Hommo Reenders, 'Het Réveil en het Heilige Land (1830–1875)', *Jaarboek voor de geschiedenis van het Nederlands protestantisme na 1800* 2 (1994) and 'Alternatieve zending'. On van de Velde and *Het Réveil* see also Fred van Lieburg: 'Een wereldwijde opwekking in het protestant vaderland. De historiografie van het Nederlands Réveil', in F. van Lieburg (ed.), *Opwekking van de natie: Het protestantse Réveil in Nederland* (Hilversum, 2012), pp. 142, 146.

240 Kirchhoff, *Text zu Land*, p. 75f.

241 Reenders, 'Het Réveil en het Heilige Land', pp. 58–9 and footnote 24, p. 70. Reenders's substantiated work is only available in Dutch and mostly ignored in international research on the subject.

242 Romans 12:6–8.

243 Van de Velde, *Narrative* II, p. 384.

244 Ibid., I, p. 71.

245 Lorraine Daston and Peter Galison, *Objectivity* (New York, 2007), pp. 39–42.

246 Claude Reignier Conder and Horatio Herbert Kitchener, *Map of Western Palestine in 26 Sheets From Surveys Conducted for the Committee of the Palestine Exploration Fund During the Years 1872–77* (London, 1880).

247 Tobler, *Bibliographia*, p. 243. Tobler to Perthes, Horn, 10 October 1865, SPA ARCH MFV 141, f. 119: 'Die karte van de Velde's zeigt einen ungeheuern fortschritt. Die schwächste seite ist die nicht immer richtige oder korrekte schreibung der ortsnamen. Dazu mangelt ihm die geduld.'

248 Haim Goren, 'Heinrich Kiepert in the Holy Land, spring 1870. Sketches from an exploration-tour of an historical cartographer', in L. Zögner (ed.), *Antike Welten. Neue Regionen. Heinrich Kiepert 1818–1899* (Berlin, 1999) and of course in this volume.

CHAPTER 3

1 Titus Tobler, *Bibliographia geographica Palaestina. Zunächst kritische Übersicht gedruckter und ungedruckter Beschreibungen der Reisen ins heilige Land* (Leipzig, 1867).
2 Reinhold Röhricht, *Bibliotheca Geographica Palaestinae. Chronologisches Verzeichnis der auf die Geographie des Heiligen Landes bezüglichen Literatur von 333 bis 1878 und Versuch einer Cartographie* (Berlin, 1890, repr. Jerusalem, 1963, ed. David H.K. Amiran), pp. 433, 646, 659.
3 Hans Fischer, 'Geschichte der Kartographie von Palästina', *ZDPV* 62 (1939) and 63 (1940), II, pp. 68, 70–71, 97.
4 Haim Goren, 'August Petermann and Palestine cartography', in S. Lentz and F. Ormeling (eds), *Die Verräumlichung des Welt-Bildes. Petermanns Geographische Mitteilungen zwischen 'explorativer Geographie' und der 'Vermessenheit' europäischer Raumphantasien* (Stuttgart, 2008); Haim Goren, *'Zieht hin und erforscht das Land'. Die deutsche Palästinaforschung im 19. Jahrhundert* (Göttingen, 2003), pp. 104–10.
5 In particular: Imre Josef Demhardt, *Der Erde ein Gesicht geben. Petermanns Geographische Mitteilungen und die Anfänge der modernen Geographie in Deutschland* (Gotha, 2006), pp. 9–26; Gerhard Engelmann, 'August Petermann als Kartographenlehrling bei Heinrich Berghaus in Potsdam', *PGM* 106 (1962); Philipp Felsch, *Wie August Petermann den Nordpol erfand* (München, 2010); Hermann Haack, 'August Petermann (1822–1878)', in H. Haack (ed.), *Schriften zur Kartographie* (Gotha and Leipzig, 1972); Matthias Hoffmann, 'August Heinrich Petermann (1822–1878). Ein Gothaer Geograph und Kartograph', *International Yearbook of Cartography* XXIX (1989); Jürgen Hagedorn, 'August Heinrich Petermann. Der erste große Wissenschaftsmanager der Geographie', *Geographie und Schule* 33/193 (2011); Rudolf Ogrissek, 'August Petermann (1822–1878). Aus seinem Leben und Wirken als Kartograph', *Vermessungstechnik* 26/11 (1978); Ewald Weller, *August Petermann. Ein Beitrag zur Geschichte der geographischen Entdeckungen und der Kartographie* (Leipzig, 1911); Hugo Wichmann, 'Petermann, August', *Allgemeine Deutsche Biographie* 26 (1888).
6 E.g. Felsch, *August Petermann*; Charles B. Officer and Jake Page, *A Fabulous Kingdom. The Exploration of the Arctic* (Oxford, 2001); David Thomas Murphy, *German Exploration of the Polar World. A History, 1870–1940* (Lincoln, 2002); Hampton Sides, *In the Kingdom of Ice. The Grand and Terrible Polar Voyage of the U.S.S. Jeanette* (New York, 2014).
7 For more detailed information, see Gerhard Engelmann, *Heinrich Berghaus. Der Kartograph von Potsdam* (Halle/Saale, 1977), and Engelmann, 'August Petermann'.
8 Gerhard Engelmann, 'Der Physikalische Atlas des Heinrich Berghaus und Alexander Keith Johnstons Physical Atlas', *PGM* 108 (1964); Jürgen Espenhorst, *Petermann's Planet. Vol. 1: The Great Handatlasses* (Schwerte, 2003).
9 Engelmann, 'August Petermann', p. 173. See also Engelmann, *Heinrich Berghaus*, pp. 108–9 and for the 1839 map Haim Goren and Bruno Schelhaas, 'Berghaus's Part of Arabia Petraea and Palestine Map (1839) and the Royal Geographical Society. Misuse or misunderstanding?', *Terrae Incognitae*, 47 (2015).
10 See Haim Goren, Chapter 1 in this book.
11 Hoffmann gives the exact date: 7 June 1845 (Hoffmann, 'August Heinrich Petermann', p. 86). See also Engelmann, 'August Petermann', pp. 181–2, and 'Physikalische Atlas', pp. 140–1.
12 James McCarthy, *Journey into Africa. The Life and Death of Keith Johnston, Scottish Cartographer and Explorer (1844–79)* (Latheronwheele, 2004), pp. 36–9; David Smith, 'The business of W. & A.K. Johnston, 1826–1901', *IMCoS Journal* 82 (2000); Diane Webster, 'Maps', in B. Bell et al. (eds), *The Edinburgh History of the Book in Scotland. Vol. 3: Ambition and Industry, 1800–1880* (Oxford, *2007*); W. & A.K. Johnston Ltd, *One hundred years of map making. The story of W. & A.K. Johnston* (Edinburgh, 1925).
13 Christopher Fleet et al., *Scotland. Mapping the Nation* (Edinburgh, 2011), p. 112.
14 Engelmann, 'Physikalische Atlas'; McCarthy, *Journey*, pp. 31–9; David Smith, 'Business', pp. 9–11.
15 Cf. his autobiography: Gustaf Kombst, *Erinnerungen aus meinem Leben* (Leipzig, 1848), but with no references to cartography, Berghaus, Petermann or Johnston.
16 McCarthy, *Journey*, p. 31.
17 John Wilson, *The Lands of the Bible, Visited and Described in an Extensive Journey Undertaken with Special Reference to the Promotion of Biblical Research and the Advancement of the Cause of Philanthropy* (2 vols, Edinburgh and London, 1847). For Wilson's biography see e.g. Mohan D. David, *John Wilson and His Institution* (Bombay, 1975); George Smith, *The Life of John Wilson, D.D. F.R.S. For Fifty Years Philanthropist and Scholar in the East* (London, 1878).
18 Probably the Indian explorer and military officer Thomas Best Jervis (1796–1857).

19 Smith, *Life of John Wilson*, p. 416.
20 Wilson, *Lands of the Bible*, pp. xii–xiv.
21 August Petermann to Justus Perthes, 10 September 1847, FBG SPA ARCH PGM 540-01-071.
22 Ulrike Kirchberger, *Aspekte deutsch-britischer Expansion. Die Überseeinteressen der deutschen Migranten in Großbritannien in der Mitte des 19. Jahrhunderts* (Stuttgart, 1999), pp. 314–25; Felsch, *August Petermann*; Richard J. Scully, '"North Sea or German Ocean"? The Anglo-German cartographic freemasonry, 1842–1914', *Imago Mundi* 62/1 (2010); Richard J. Scully, *British Images of Germany. Admiration, Antagonism & Ambivalence, 1860–1914* (Basingstoke, 2012), pp. 30–6.
23 Scully, 'North Sea'.
24 Francis Herbert, 'The Royal Geographical Society's membership, the map trade, and geographical publishing in Britain 1830 to ca 1930. An introductory essay with listing of some 250 fellows in related professions', *Imago Mundi* 35 (1983), p. 71. See also: Felsch, *August Petermann*, pp. 43–51; McCarthy, *Journey*, pp. 38–9; David Smith, 'Business', pp. 11–13.
25 For Bunsen's biography see e.g. Frank Foerster, *Christian Carl Josias Bunsen. Diplomat, Mäzen und Vordenker in Wissenschaft, Kirche und Politik* (Bad Arolsen, 2001); Erich Geldbach (ed.), *Der Gelehrte Diplomat. Zum Wirken Christian Carl Josias Bunsens* (Leiden, 1980); Hans-Rudolf Ruppel, *Universeller Geist und guter Europäer. Christian Carl Josias von Bunsen 1791–1860. Beiträge zu Leben und Werk des 'Gelehrten Diplomaten' zum 200. Geburtstag* (Korbach, 1991).
26 Kurt Schmidt-Clausen, 'Der Beitrag Bunsens zur Gründung des Bistums Jerusalem', in E. Geldbach (ed.), *Der Gelehrte Diplomat. Zum Wirken Christian Carl Josias Bunsens* (Leiden, 1980).
27 See Haim Goren, Chapter 1 in this volume.
28 See Jutta Faehndrich, Chapter 2 in this volume.
29 August Petermann, 'On the fall of the Jordan, and of principal rivers in the United Kingdom', *JRGS* 18 (1848); Edward Robinson, 'Depression of the Dead Sea and of the Jordan Valley', *JRGS* 18 (1848); Cf. Haim Goren, *Dead Sea Level. Science, Exploration and Imperial Interests in the Near East* (London, 2011), pp. 249–61; Haim Goren and Bruno Schelhaas, 'August Petermann vergleicht Flussgefälle', in P. Weigel et al. (eds), *Das Heilige Land in Gotha. Der Verlag Justus Perthes und die Palästinakartographie im 19. Jahrhundert* (Gotha, 2014); Weller, *August Petermann*, pp. 114–15.
30 For RGS history see in particular Felix Driver, *Geography Militant. Cultures of Exploration and Empire* (Oxford, 2001).
31 Herbert, 'The Royal Geographical Society's membership', p. 87.
32 *JRGS* 16 (1846), p. xxii–xxx.
33 *JRGS* 21 (1851), p. 1; 22 (1852), p. lii; 23 (1853), p. liii.
34 Herbert, 'The Royal Geographical Society's membership', p. 87.
35 Sondra Miley Cooney, 'William Somerville Orr, London publisher and printer. The skeleton in W. & R. Chamber's closet', in J. Hinks and C. Armstrong (eds), *Worlds of Print. Diversity in the Book Trade* (New Castle, DE, 2006).
36 For Petermann's London years see Felsch, *August Petermann*, pp. 63–129.
37 Cf. Arthur H. Robinson, *Early Thematic Mapping in the History of Cartography* (Chicago and London, 1982), passim.
38 Louis von Wildenbruch, 'Notes on the physical geography of Palestine', *JRGS* 20 (1850).
39 August Petermann, 'Note, containing tabular summary, &c.', *JRGS* 20 (1850).
40 Robinson, 'Depression'.
41 Petermann, 'On the fall of the Jordan'.
42 Thomas Howard Molyneux, 'Expedition to the Jordan and the Dead Sea', *JRGS* 18 (1848); Goren, *Dead Sea Level*, pp. 249–61.
43 August Petermann, (Notes on Palestine), FBG SPB Konvolut 3030 Syrien I. The English with German plug-ins is Petermann's.
44 For Symonds see Goren, 'British surveyors in Palestine and Syria, 1840–1841', International Cartographic Conference A Coruña 2005: Mapping approaches into a changing world, available at http://icaci.org/documents/ICC_proceedings/ICC2005/htm/pdf/oral/TEMA16/Session%205/HAIM%20GOREN.pdf (accessed 13 August 2016); Goren, *Dead Sea Level*, pp. 102–5, 231–7, passim; Yolande Jones, 'British military surveys of Palestine and Syria 1840–1841', *The Cartographic Journal* 10/1 (1973).
45 For Wildenbruch see Goren, *Zieht hin*, pp. 181–5 and Goren, *Dead Sea Level*, pp. 245–7. See also the correspondence of Ernestine von Wildenbruch in Albrecht von Boguslawski (ed.), *Aus der preussischen Hof- und diplomatischen Gesellschaft* (Stuttgart and Berlin, 1903).

46 Petra Weigel, *Sammlung Perthes Gotha* (Berlin, 2011).
47 Schelhaas, '"Das Wiederkehren des Fragezeichens in der Karte". Gothaer Kartenproduktion im 19. Jahrhundert', *Geographische Zeitschrift* 97/1 (2009); Bruno Schelhaas and Ute Wardenga, '"Inzwischen spricht die Karte für sich selbst". Transformation von Wissen im Prozess der Kartenproduktion', in S. Siegel and P. Weigel (eds), *Die Werkstatt des Kartographen. Materialien und Praktiken visueller Welterzeugung* (München, 2011).
48 August Petermann, 'Bemerkungen über die Depression des Todten Meeres', London, 14 December 1847, FBG SPB Konvolut 3030 Syrien I.
49 Wilhelm Mahlmann, 'Russeggers Bemerkungen über dessen meteorologische Hauptdurchschnitte und die barometrische Höhenbestimmung des Todten Meeres', *MVGEB* N.F. 3 (1846); Joseph von Russegger, *Reisen in Europa, Asien und Afrika. Vol. 3: Reise in Unter-Egypten, auf der Halbinsel des Sinai und im gelobten Lande* (Stuttgart, 1847). For Russegger see Goren, *Dead Sea Level*, pp. 217–20.
50 Carl Ritter, 'Wildenbruch's Bemerkungen über Gaillardot's Karte', *MVGEB* N.F. 3 (1846); Carl Ritter, 'L. v. Wildenbruch's Nivellements in Syrien', *MVGEB* N.F. 3 (1846).
51 For their biographies and roles within the debate, cf. Goren, *Zieht hin*, passim.
52 Carl Ritter to August Petermann, 21 April 1849, FBG SPA ARCH PGM 473-003-05.
53 August Petermann to Carl Ritter, 1 June 1849, SBB-PK Nachl. Carl Ritter, K 4, p. 135.
54 Carl Ritter to Edward Robinson, 25 September 1852, ERP, ER, L. 4 Ritter.
55 FBG SPK 547_112553079. See also Haim Goren, 'Lieutenant Symonds irrt bei Höhenmessungen', in P. Weigel et al. (eds), *Das Heilige Land in Gotha. Der Verlag Justus Perthes und die Palästinakartographie im 19. Jahrhundert* (Gotha, 2014).
56 Petermann, 'On the fall of the Jordan', p. 104; *JRGS* 19 (1849), p. ciii.
57 John Relly Beard, *The People's Dictionary of the Bible* (2 vols, London, first edn 1847/8).
58 Charles Gordon Bolam, 'John Relly Beard and the New Function of Unitarianism in the 1850s', *Transactions of the Unitarian Historical Society* 12/1 (1959); Geoffrey Head, *The Life & Times of John Relly Beard (1800–1876)* (Bolton, 1997); Herbert McLachlan, *Records of a Family 1800–1933. Pioneers in Education, Social Service and Liberal Religion* (Manchester, 1935).
59 Georg Benedict Winer, *Biblisches Realwörterbuch zum Handgebrauch für Studirende, Kandidaten, Gymnasiallehrer und Prediger* (Leipzig, 1820); John Kitto, *A Cyclopaedia of Biblical Literature* (2 vols, Edinburgh, 1845).
60 Beard, *People's Dictionary* 1847/8, vol. I, p. vi.
61 For the history of Simpkin, Marshall & Co., Henry Curwen, *A History of Booksellers. The Old and New* (London, 1873), pp. 412–20. The cover page lists also as cooperating publishers: Ainsworth (Manchester), Wiley & Putnam (New York) and Crosby & Nicholls (Boston).
62 National Library of Scotland (NLS), NF.725.b.21.
63 Gotthilf Heinrich von Schubert, *Reise in das Morgenland in den Jahren 1836 und 1837*, I–III (Erlangen, 1839); Goren, *Zieht hin*, pp. 117–24; Goren, *Dead Sea Level*, pp. 207–11. For Schubert's biography see Andreas Eichler, *G. H. Schubert. Ein anderer Humboldt* (Niederfrohna, 2010); Gotthilf Heinrich von Schubert, *Der Erwerb aus einem vergangenen und die Erwartungen von einem zukuenftigen Leben. Eine Selbstbiographie* (5 vols, Erlangen, 1854–6).
64 The exact location of the biblical Kadesh-Barnea was a research question for long time. Cf. for example H. Clay Trumbull, *Kadesh-Barnea, its Importance and Probable Site with the Story of a Hunt for it Including Studies of the Route of the Exodus and the Southern Boundary of the Holy Land* (New York, 1884), and Rudolph Cohen, *Kadesh-Barnea. A Fortress from the Time of the Judean Kingdom* (Jerusalem, 1983).
65 John R. Beard to August Petermann, 1 September 1848, FBG SPB Konvolut 3030 Syrien I.
66 Beard, *People's Dictionary* 1850 and 1861.
67 FBG SPK 547_112549810.
68 Molyneux, 'Expedition to the Jordan'; Goren, *Dead Sea Level*, pp. 249–61; Haim Goren, 'Nicholayson and Finn describe the expeditions and death of Costigan and Molyneux', *Cathedra* 85 (1997).
69 August Petermann to Carl Ritter, 1 June 1849, SBB-PK Nachl. Carl Ritter, K 4, p. 135.
70 FBG SPK 547_112619053.
71 *JRGS* 19 (1849), p. cii.
72 John Relly Beard, *Scripture Illustrated from Recent Discoveries in the Geography of Palestine* (London, 1849); *Scripture Vindicated Against Some Perversions of Rationalism* (London, 1849); and *A Biblical Reading Book for Schools and Families* (London, 1849).
73 John Relly Beard, *A Biblical Atlas. With a Brief Geographical Introduction, and a Complete Scriptural Gazetteer* (London, 1849).

74 *The Christian Reformer*, New Series VI (1850), p. 51.
75 Jeremy Black, *Maps and History. Constructing Images of the Past* (New Haven, 1997), pp. 32–5.
76 William Hughes, *The Illuminated Atlas of Scripture Geography* (London, 1840); Society for Diffusion of Useful Knowledge, *Maps of the Society for the Diffusion of Useful Knowledge.* Vol. 1 (London, 1844).
77 Thomas Milner and August Petermann, *The Atlas of Physical Geography* (London, 1850); Cf. Bruno Schelhaas, 'August Petermann visualisiert Höhen- und Klimadaten' in P. Weigel et al. (eds), *Das Heilige Land in Gotha. Der Verlag Justus Perthes und die Palästinakartographie im 19. Jahrhundert* (Gotha, 2014).
78 James A. Secord, *Victorian Sensation. The Extraordinary Publication, Reception and Secret Authorship of 'Vestiges of the Natural History of Creation'* (Chicago, 2000), pp. 463–4.
79 Milner and Petermann, *Atlas of Physical Geography*, p. 4.
80 Espenhorst, *Petermann's Planet*, pp. 385–6. See also Engelmann, 'Physikalische Atlas', p. 141, which presents a slightly different edition history of Petermann's first atlas.
81 The *Spectator*, 19 January 1850, p. 64.
82 Milner and Petermann, 'Preface', *Atlas of Physical Geography.*
83 Ibid., pp. 104–6.
84 Ibid., pp. 135–6.
85 Petermann, 'Note', p. 232.
86 August Petermann, 'Climatological History of the Months in Palestine', FBG SPB Konvolut Palästina I.
87 August Petermann, 'Vegetation of Palestine', FBG SPB Konvolut Palästina I.
88 For Kitto's biography see Jonathan Edwards Ryland, *Memoirs of John Kitto, Editor of 'the Pictorial Bible' … Compiled Chiefly from his Letters and Journals, with a Critical Estimate of Dr. Kitto's Life and Writings by Prof. Eadie* (Edinburgh, 1856); John Eadie, *Life of John Kitto, D.D., F.S.A.* (Edinburgh, 1861).
89 Ludwig von Wildenbruch to August Petermann, 17 April 1849, FBG SPB Konvolut Syrien I.
90 FBG SPK 547_112620078; Cf. Schelhaas, 'August Petermann verwertet' Kartenentwürfe. In P. Weigel et al. (eds), Das Heilige Land in Gotha. *Der Verlag Justus Perthes und die Palästinakartographie im 19. Jahrhundert* (Gotha, 2014).
91 *JRGS* 19 (1849), p. ciii.
92 Thomas Milner and August Petermann, *A Descriptive Atlas of Astronomy, and of Physical and Political Geography* (London 1850) p. 154.
93 Thomas Milner and August Petermann, *The Atlas of Political Geography* (London, 1851), p. 50.
94 Thomas Milner and August Petermann, *The Library Atlas of Physical and Political Geography* (London, 1854).
95 John Dower, *A New General Atlas of Modern Geography*, 2nd edn (London, 1860); John Dower, *School Atlas of Modern Geogrpahy* (London, 1852).
96 In particular, Demhardt, *Erde*; Jan Smits, *Petermann's Maps. Carto-Bibliography of the Maps in Petermanns Geographische Mitteilungen 1855–1945* ('t Goy-Houten, 2004).
97 For Roth see Goren, *Zieht hin*, pp. 124–39; for Tobler see Heinrich Jakob Heim, *Dr. Titus Tobler der Palästinafahrer. Ein appenzellisches Lebensbild nach handschriftlichen Quellen* (Zürich and Trogen, 1879); Robert Hinderlig and Stefan Sonderegger, 'Titus Tobler (1806–1877). Ein appenzellisches Universalgenie. Neue Ergebnisse aus Nachlassfunden', *Appenzellische Jahrbücher* 121 (1993); for van de Velde see Jutta Faehndrich, Chapter 2 in this book.
98 Bruno Schelhaas, 'August Petermann sammelt Höhenmessungen' in P. Weigel et al. (eds), *Das Heilige Land in Gotha. Der Verlag Justus Perthes und die Palästinakartographie im 19. Jahrhundert* (Gotha, 2014).
99 August Petermann, 'Die projektirte Kanalisirung des Isthmus von Sues, nebst Andeutungen über die Höhen-Verhältnisse der angrenzenden Regionen, besonders Palästina', *PGM* 1 (1855).
100 Cf. Hubert Bonin, *History of the Suez Canal Company* (Genève, 2010).
101 Petermann, 'Kanalisirung', p. 364.
102 Ibid., p. 366.
103 Schelhaas, 'Wiederkehren'; Schelhaas and Wardenga, 'Inzwischen spricht'.
104 William Allen, *The Dead Sea. A New Route to India* (London, 1855); Cf. Goren, *Dead Sea Level*, pp. 119–21.
105 Josias Leslie Porter, *Five Years in Damascus* (2 vols, London, 1855). Cf. Edwin James Aiken, *Scriptural Geography. Portraying the Holy Land* (London, 2010), pp. 89–132.
106 Petermann, 'Kanalisirung', p. 375.
107 Haim Goren and Bruno Schelhaas, 'Conrad Schick plant Straßen und Eisenbahnen', in P. Weigel et al. (eds), *Das Heilige Land in Gotha. Der Verlag Justus Perthes und die Palästinakartographie im 19. Jahrhundert* (Gotha, 2014).

108 Conrad Schick, 'Studien über Strassen- und Eisenbahn-Anlagen zwischen Jaffa und Jerusalem, nebst Beschreibung der Gegend im Norden von Jaffa und der Ruinen von Mar Zacharias', *PGM* 13 (1867).
109 Charles Franz Zimpel, *Straßen-Verbindung des Mittelländischen mit dem Todten Meere und Damascus über Jerusalem ...* (Frankfurt am Main, 1865). Cf. Paul Cotterell, *The Railways of Palestine and Israel* (Abingdon, 1984).
110 Alex Carmel, 'Wie es zu Conrad Schicks Sendung nach Jerusalem kam', *ZDPV* 99 (1983); Goren and Schelhaas, 'Conrad Schick plant Straßen'; Strobel, *Schick*.
111 FBG SPK 547_112550002.
112 FBG SPK 547_112550002.
113 Conrad Schick to Justus Perthes, 12 October 1866, FBG SPB Konvolut Palästina 29, I, p. 21.
114 Schick, 'Studien über Straßen', pp. 125 and 129.
115 For further details, see Espenhorst, *Petermann's Planet*, pp. 179–449.
116 Petra Weigel, 'Karl Georg von Raumer betreibt Bibelkartographie', in P. Weigel et al. (eds), *Das Heilige Land in Gotha. Der Verlag Justus Perthes und die Palästinakartographie im 19. Jahrhundert* (Gotha, 2014).
117 Adolf Stieler, *Hand-Atlass über alle Theile der Erde* (Gotha, 1817–1945) 2nd edn 1851, p. 15.
118 Karl von Raumer, *Palästina* (Leipzig, 1838); Karl von Raumer, *Beiträge zur biblischen Geographie. Nebst einem Höhendurchschnitt* (Leipzig, 1843).
119 Petra Weigel, 'August Petermann modernisiert alte Kartenbilder' in P. Weigel et al. (eds), *Das Heilige Land in Gotha. Der Verlag Justus Perthes und die Palästinakartographie im 19. Jahrhundert* (Gotha, 2014).
120 Petermann, 'Vollendung der Jubel-Ausgabe von Stieler's Hand-Atlas in 84 Blättern. Notizen zur Kartographie Palästina's und Polynesiens', *PGM* 14 (1868), p. 373.

Bibliography

ARCHIVAL SOURCES

Staatsbibliothek zu Berlin – Preußischer Kulturbesitz, Handschriftenabteilung (SBB-PK)

Carl Ritter Papers	SBB-PK, Nachl. Carl Ritter
Sammlung Darmstaedter	SBB-PK, Slg. Darm. Asien

Staatsbibliothek zu Berlin – Preußischer Kulturbesitz, Kartenabteilung (SBB-PK)

Map Department	SBB-PK Kart

Universität Erfurt, Forschungsbibliothek Gotha, Sammlung Perthes Gotha (FBG SP)

Sammlung Perthes Archiv, Schriftleitung Petermanns Geographische Mitteilungen	FBG SPA ARCH PGM 141; 473; 540
Sammlung Perthes Archiv, Mitarbeiter und Freunde des Verlages	FBG SPA ARCH MFV 9; 51
Sammlung Perthes Archiv, Firmen- und Familienarchiv, Verlagsauslieferungsbücher	FBG SPA ARCH FFA VAB 1; 2
Sammlung Perthes, Verlagsbibliothek, Konvolut Palästina	FBG SPB Konvolut As 29
Sammlung Perthes, Verlagsbibliothek, Konvolut Syrien	FBG SPB Konvolut 3030
Sammlung Perthes, Kartensammlung	FBG SPK

Universitätsbibliothek Freiburg (UBF)

Carl Ritter Papers UBF, Ritter NL 33

Studienzentrum August Hermann Francke, Archiv (Francke Foundations, Archives) (AFSt), Halle/Saale

Friedrich August Gottreu Tholuck Papers AFSt/N, THOLUCK

Royal Geographical Society, Collections, London

RGS-IBG Collections

Hamilton College, Burke Library, Special Collections, Clinton, NY

Edward Robinson Papers ERP

Harvard University, Houghton Library, Cambridge, MA

Eli Smith Papers, 1819–1869 (ABC 60) ESP

PERIODICALS

BS	*Bibliotheca Sacra: or Tracts and Essays on Topics Connected with Biblical Literature and Theology*
BSTR	*Bibliotheca Sacra and Theological Review*
JRGS	*Journal of the Royal Geographical Society*
MVGEB	*Monatsberichte über die Verhandlungen der Gesellschaft für Erdkunde zu Berlin*
PGM	*Petermanns Geographische Mitteilungen (1855–78: Mittheilungen aus Justus Perthes' Geographischer Anstalt über wichtige neue Erforschungen auf dem Gesammtgebiete der Geographie)*
ZDPV	*Zeitschrift des deutschen Palästina-Vereins*

BOOKS AND ARTICLES

Anon., 'Geography and Biography of the Old Testament', *The London Quarterly Review* (American edn) 106 (October 1859), pp. 203–31.

Anon., 'Obituary: C. W. M. Van de Velde', *The Times*, 24 March 1898, p. 10.

Aiken, Edwin James, *Scriptural Geography. Portraying the Holy Land* (London, 2010).
Allen, William, *The Dead Sea. A New Route to India* (London, 1855).
Anker, Valentina, 'Calame, Alexandre', Historisches Lexikon der Schweiz (HLS), available at http://www.hls-dhs-dss.ch/textes/d/D21996.php (version of 14 July 2003, accessed 23 June 2015).
Arnold, Friedrich August, 'Map of the Holy Land (review)', *Neue Jahrbücher für Philologie und Pädagogik* 80 (1859), pp. 251–6.
Ballenthin, Sven, 'Commander Mansell vermisst Küsten und Meerestiefen', in P. Weigel et al. (eds), *Das Heilige Land in Gotha. Der Verlag Justus Perthes und die Palästinakartographie im 19. Jahrhundert* (Gotha, 2014), pp. 76–7.
Beard, John Relly, *A Biblical Atlas. With a Brief Geographical Introduction, and a Complete Scriptural Gazetteer* (London, 1849).
———, *A Biblical Reading Book for Schools and Families* (London, 1849).
———, *The People's Dictionary of the Bible* (2 vols, London, 1847/8, 1850 (3rd edn), repr. Manchester, 1861).
———, *Scripture Illustrated from Recent Discoveries in the Geography of Palestine* (London, 1849).
———, *Scripture Vindicated Against Some Perversions of Rationalism* (London, 1849).
Ben-Arieh, Yehoshua, 'The first surveyed maps of Jerusalem', *Eretz-Israel. Archaeological, Historical and Geographical Studies* 11 (1973), pp. 64–74. [Heb.]
———, 'Holy Land views in 19th century western travel literature', in M. Davis and Y. Ben-Arieh (eds), *Western Societies in the Holy Land* (Westport, 1991), pp. 10–29.
———, 'Perceptions and images of the Holy Land', in R. Kark (ed.), *The Land that Became Israel. Studies in Historical Geography* (New York, London, Jerusalem, 1989), pp. 37–53.
———, 'Pioneer scientific exploration in the Holy Land at the beginning of the nineteenth century', *Terrae Incognitae* 4 (1972), pp. 95–110.
———, *The Rediscovery of the Holy Land in the 19th Century* (Jerusalem, 1979, 1983 (2nd edn)).
Benvenisti, Meron, *Sacred Landscape. The Buried History of the Holy Land since 1948* (Berkeley, 2000).
Berghaus, Heinrich, *Geographisches Memoir zur Erklärung und Erläuterung der Karte von Syrien* (Gotha, 1835).
———, *Geo-hydrographisches Memoir zur Erklärung und Erläuterung der General-Karte von Arabia und dem Nil-Lande* (Gotha, 1835).
———, *Karte von Syrien, den Manen Jacotin's und Burckhardt's gewidmet* (Gotha, 1835).
———, *Physikalischer Atlas. Oder Sammlung von Karten, auf denen die hauptsächlichen Erscheinungen der anorganischen und organischen Natur nach ihrer geographischen Verbreitung und Vertheilung bildlich dargestellt sind* (Gotha, 1838–48).
———, 'Über die Reisen von E. Robinson und E. Smith im Peträischen Arabien und im Gelobten Lande', *Almanach für das Jahr 1840. Der Belehrung und Unterhaltung auf dem Gebiete der Erd-, Länder-, Völker- und Staatenkunde* 4 (1840), pp. 531–5.
Bertou, Jules de, 'Extract from a notice on the site of ancient Tyre', *JRGS* 9 (1839), pp. 286–94.
———, 'Notes on a journey from Jerusalem to Hebron, the Dead Sea, El Ghór, and Wádi 'Arabah to 'Akabah, and back by Petra, in April, 1838', *JRGS* 9 (1839), pp. 277–86.
Bird, Isaac, *Bible Work in Bible Lands, or, Events in the History of the Syria Mission* (Philadelphia, 1872).
Black, Jeremy, *Maps and History. Constructing Images of the Past* (New Haven, 1997).
Bliss, Frederick Jones, *The Development of Palestine Exploration* (London and New York, 1906).
Blondel, Lucien A., *Carte du Liban d'après les reconnaissances de la brigade topographique du corps expéditionnaire de Syrie en 1860–1861, dressée au Dépôt de la Guerre, étant directeur le General Blondel, sous le ministère de S. E. le Marechal Comte Randon* (Paris, 1862).
Blow Williams, Judith, *British Commercial Policy and Trade Expansion 1750–1850* (Oxford, 1972).

Boguslawski, Albrecht von (ed.), *Aus der preussischen Hof- und diplomatischen Gesellschaft* (Stuttgart and Berlin, 1903).
Bolam, Charles Gordon, 'John Relly Beard and the new function of Unitarianism in the 1850s', *Transactions of the Unitarian Historical Society* 12/1 (1959), pp. 155–62.
Bonato, Lucie, 'Camille Callier. Un officier instruit de l'armée française qui explora Chypre en 1832', *Thetis* 10 (2003), pp. 113–41.
Bonin, Hubert, *History of the Suez Canal Company* (Genève, 2010).
Bourdaloue, Paul-Adrien, *Isthme de Suez. Parallèle entre le nivellement de 1799 at celui de 1847* (Bourges, 1856)
Brogiato, Heinz Peter, 'Gotha als Wissens-Raum', in S. Lentz and F. Ormeling (eds), *Die Verräumlichung des Welt-Bildes. Petermanns Geographische Mitteilungen zwischen 'explorativer Geographie' und der 'Vermessenheit' europäischer Raumphantasien* (Stuttgart, 2008), pp. 15–39.
Brown, Jerry Wayne, *The Rise of Biblical Criticism in America, 1800–1870. The New England Scholars* (Middletown, CT, 1969).
Buckingham, James Silk, *Travels in Palestine. Through the Countries of Bashan and Gilead, East of the River Jordan, Including a Visit to the Cities of Geraza and Gamala in the Decapolis* (2 vols, London, 1822).
Burckhardt, John Lewis, *Travels in Syria and the Holy Land* (London, 1822).
Butlin, Robin, 'George Adam Smith and the historical geography of the Holy Land. Contents, contexts and connections', *Journal of Historical Geography* 14/4 (1988), pp. 381–401.
Callier, Camille, *Carte de la Syrie Méridionale et de la Palestine, dressée, en 1835, d'après les ordres du Directeur du Dépôt-général de la guerre Lieut.-Général Pelet* ([Paris], 1840).
———, 'Voyage en Asie Mineure, en Syrie, en Palestine et en Arabie-Pétrée', *Bulletin de la Société de Géographie de Paris* 7 (1834), pp. 5–22, 239–62.
Carmel, Alex, 'C. F. Spittler and the activities of the pilgrims mission in Jerusalem', in Gad G. Gilbar (ed.), *Ottoman Palestine 1880–1914: Studies in Economic and Social History* (Leiden, 1990), pp. 255–86.
———, 'Wie es zu Conrad Schicks Sendung nach Jerusalem kam', *ZDPV* 99 (1983), pp. 204–18.
Cohen, Rudolph, *Kadesh-Barnea. A Fortress from the Time of the Judean Kingdom* (Jerusalem, 1983).
Colton, Joseph Hutchins, *New Map of Palestine from the Latest Authorities. Chiefly from the Maps and Drawings of Robinson and Smith, with Corrections and Additions Furnished by the Rev. Dr. Robinson* (New York, 1845).
Conder, Claude Reignier and Horatio Herbert Kitchener, *Map of Western Palestine in 26 Sheets From Surveys Conducted for the Committee of the Palestine Exploration Fund During the Years 1872–77* (London, 1880).
Cooney, Sondra Miley, 'William Somerville Orr, London publisher and printer. The skeleton in W. & R. Chamber's closet', in J. Hinks and C. Armstrong (eds), *Worlds of Print. Diversity in the Book Trade* (New Castle, DE, 2006), pp. 135–48.
Cotterell, Paul, *The Railways of Palestine and Israel* (Abingdon, 1984).
Curwen, Henry, *A History of Booksellers. The Old and New* (London, 1873).
Daston, Lorraine and Peter Galison, *Objectivity* (New York, 2007).
David, Mohan D., *John Wilson and His Institution* (Bombay, 1975).
Davis, John, *The Landscape of Belief. Encountering the Holy Land in Nineteenth-Century American Art and Culture* (Princeton, NJ, 1996).
Dawson, Llewellyn Stiles, *Memoirs of Hydrography. Including Brief Biographies of the Principal Officers Who Have Served in H.M. Naval Surveying Service Between the Years 1750 and 1855*, Vol. II: *1830 to 1885* (Eastbourne, 1885).

Dawson, Warren R. and Eric P. Uphill, *Who Was Who in Egyptology*, ed. M. L. Bierbrier (London, 1995 (3rd edn)).
Day, Archibald, *The Admiralty Hydrographic Service, 1795–1919* (London, 1967).
Demhardt, Imre Josef, *Der Erde ein Gesicht geben. Petermanns Geographische Mitteilungen und die Anfänge der modernen Geographie in Deutschland* (Gotha, 2006).
Doergens, Richard, 'Barometrische Höhenbestimmungen in Syrien und Palästina, ausgeführt im Jahre 1860', *PGM* 12 (1866), pp. 97–102.
Dower, John, *School Atlas of Modern geography* (London, 1852).
—— *A New General Atlas of Modern Geography* (London, 1860 (2nd edn)).
Driver, Felix, *Geography Militant. Cultures of Exploration and Empire* (Oxford, 2001).
Dufour, Guillaume-Henri, *Topographische Karte der Schweiz, Carte topographique de la Suisse, Carta topografica della Svizzera* (Geneva, 1845–65).
Eadie, John, *Life of John Kitto, D.D., F.S.A.* (Edinburgh, 1861).
Edney, Matthew H., *Mapping an Empire. The Geographical Construction of British India, 1765–1843* (Chicago, 1997).
Ehrenberg, Christian Gottfried, *Die Corallenthiere des rothen Meeres, physiologisch untersucht und systematisch verzeichnet* (Berlin, 1834).
Eichler, Andreas, *G. H. Schubert. Ein anderer Humboldt* (Niederfrohna, 2010).
Eißfeldt, Otto, 'Wilhelm Gesenius und die Palästina-Wissenschaft. Zur 100. Wiederkehr seines Todestages († 23. Oktober 1842)', *ZDPV* 65 (1942), pp. 8–112.
Engelmann, Gerhard, 'August Petermann als Kartographenlehrling bei Heinrich Berghaus in Potsdam', *PGM* 106 (1962), pp. 161–82.
———, *Heinrich Berghaus. Der Kartograph von Potsdam* (Halle/Saale, 1977).
———, 'Der Physikalische Atlas des Heinrich Berghaus und Alexander Keith Johnstons Physical Atlas', *PGM* 108 (1964), 133–49.
Engelmann, Gerhard and Egon Klemp, 'Carl Ritters Briefsammlung in der Handschriftenabteilung/ Literaturarchiv der Deutschen Staatsbibliothek', in H. Richter (ed.), *Carl Ritter. Werk und Wirkungen* (Gotha, 1983), pp. 245–54.
Espenhorst, Jürgen, *Petermann's Planet.* Vol. 1: *The Great Handatlasses* (Schwerte, 2003).
Ewald, Heinrich von, 'Übersicht der 1859–1860 erschienenen Schriften', *Jahrbücher der Biblischen Wissenschaft* 10 (1860), pp. 124–294.
Faehndrich, Jutta, 'Charles William Meredith van de Velde entwirf mehr als ein Kartenbild', in P. Weigel et al. (eds), *Das Heilige Land in Gotha. Der Verlag Justus Perthes und die Palästinakartographie im 19. Jahrhundert* (Gotha, 2014), pp. 104–5.
———, 'Charles William Meredith van de Velde schreibt über das Kartenmachen', in P. Weigel et al. (eds), *Das Heilige Land in Gotha. Der Verlag Justus Perthes und die Palästinakartographie im 19. Jahrhundert* (Gotha, 2014), pp. 84–5.
———, 'Charles William Meredith van de Velde zeichnet biblische Geschichte', in P. Weigel et al. (eds), *Das Heilige Land in Gotha. Der Verlag Justus Perthes und die Palästinakartographie im 19. Jahrhundert* (Gotha, 2014), pp. 89–91.
———, 'Theodor Menke konzentriert Bibelkarten', in P. Weigel et al. (eds), *Das Heilige Land in Gotha. Der Verlag Justus Perthes und die Palästinakartographie im 19. Jahrhundert* (Gotha, 2014), pp. 128–9.
———, 'Titus Tobler sammelt Straßennamen', in P. Weigel et al. (eds), *Das Heilige Land in Gotha. Der Verlag Justus Perthes und die Palästinakartographie im 19. Jahrhundert* (Gotha, 2014), pp. 106–7.
Farah, Caesar E., *The Politics of Interventionism in Ottoman Lebanon, 1830–1861* (Oxford, London and New York, 2000).
Felsch, Philipp, *Wie August Petermann den Nordpol erfand* (München, 2010).

Fischer, Hans, 'Geschichte der Kartographie von Palästina', *ZDPV* 62 (1939), pp. 169–89 and 63 (1940), pp. 1–111.

Fleet, Christopher, Margaret Wilkes and Charles W.J. Withers, *Scotland. Mapping the Nation* (Edinburgh, 2011).

Foerster, Frank, *Christian Carl Josias Bunsen. Diplomat, Mäzen und Vordenker in Wissenschaft, Kirche und Politik* (Bad Arolsen, 2001).

Freeman, Thomas Walter, 'Carl Ritter und die englische Geographie', in H. Richter (ed.), *Carl Ritter. Werk und Wirkungen* (Gotha, 1983), pp. 93–102.

Freitag, Ulrich, 'Heinrich Kieperts kartographische Leistung', in L. Zögner (ed.), *Antike Welten. Neue Regionen. Heinrich Kiepert 1818–1899* (Berlin, 1999), pp. 23–30.

Galazka, Guy, *A la découverte de la Palestine. Voyageurs français en Terre sainte au XIXe siècle* (Paris, 2011).

Gelber, Nathan Michael, 'A pre-Zionist plan for colonizing Palestine. The proposal of a non-Jewish German-American in 1853' (trsl. Guido Kisch), *Historia Judaica*, 1/2 (1939), pp. 81–90.

Geldbach, Erich (ed.), *Der Gelehrte Diplomat. Zum Wirken Christian Carl Josias Bunsens* (Leiden, 1980).

Gidney, William Thomas, *The History of the London Society for Promoting Christianity Amongst the Jews. From 1809 to 1908* (London, 1908).

Giltner, John H., *Moses Stuart. The Father of Biblical Science in America* (Atlanta, 1988).

Glass, Dagmar, *Malta, Beirut, Leipzig and Beirut again. Eli Smith, the American Syria Mission and the Spread of Arabic Typography in 19th Century Lebanon* (Beirut, 1998).

Godlewska, Anne, *Geography Unbound. French Geographical Science from Cassini to Humboldt* (Chicago, 1999)

———, 'Napoleon's geographers (1797–1815). Imperialists and soldiers of modernity', in A. Godlewska and N. Smith (eds), *Geography and Empire* (Oxford and Cambridge, MA, 1994), pp. 34–53.

———, *The Napoleonic Survey of Egypt. A Masterpiece of Cartographic Compilation and Early Nineteenth-Century Fieldwork* (Toronto, 1988).

Goren, Haim, 'August Petermann and Palestine cartography', in S. Lentz and F. Ormeling (eds), *Die Verräumlichung des Welt-Bildes. Petermanns Geographische Mitteilungen zwischen 'explorativer Geographie' und der 'Vermessenheit' europäischer Raumphantasien* (Stuttgart, 2008), pp. 75–87.

———, 'British surveyors in Palestine and Syria, 1840–1841', International Cartographic Conference A Coruña 2005: *Mapping Approaches into a Changing World*, http://icaci.org/documents/ICC_proceedings/ICC2005/htm/pdf/oral/TEMA16/Session%205/HAIM%20GOREN.pdf (accessed 13 August 2016).

———, 'Carl Ritter ermutigt junge Kartographen', in P. Weigel et al. (eds), *Das Heilige Land in Gotha. Der Verlag Justus Perthes und die Palästinakartographie im 19. Jahrhundert* (Gotha, 2014), pp. 42–3.

———, 'Carl Ritter's contribution to Holy Land research', in A. Buttimer, S. D. Brunn and U. Wardenga (eds), *Text and Image. Social Construction of Regional Knowledge* (Leipzig, 1999), pp. 28–37.

———, *Dead Sea Level. Science, Exploration and Imperial Interests in the Near East* (London, 2011).

———, 'Ernst Gustav Schultz referiert über Jerusalem', in P. Weigel et al. (eds), *Das Heilige Land in Gotha. Der Verlag Justus Perthes und die Palästinakartographie im 19. Jahrhundert* (Gotha, 2014), pp. 100–1.

———, 'Heinrich Kiepert in the Holy Land, spring 1870. Sketches from an exploration-tour of an historical cartographer', in L. Zögner (ed.), *Antike Welten. Neue Regionen. Heinrich Kiepert 1818–1899* (Berlin, 1999), pp. 45–62.

———, 'Lieutenant Symonds irrt bei Höhenmessungen', in P. Weigel et al. (eds), *Das Heilige Land in Gotha. Der Verlag Justus Perthes und die Palästinakartographie im 19. Jahrhundert* (Gotha, 2014), pp. 70–71.

———, 'Nicholayson and Finn describe the expeditions and death of Costigan and Molyneux', *Cathedra* 85 (1997), pp. 65–95. [Heb.]

———, 'Pioneer surveying and mapping of the Galilee and the Hauran. The Map of the Sources of the Jordan River (1841) and Map of the *Ledja* (1838)', *Cathedra* 96 (July 2000), pp. 111–30 [Heb.]

———, 'Sacred, but not surveyed. 19th century surveys of Palestine', *Imago mundi* 54 (2002), pp. 87–110.

———, 'Scientific organizations as agents of change. The Palestine Exploration Fund, the Deutsche Verein zur Erforschung Palästinas and nineteenth-century Palestine', *Journal of Historical Geography* 27 (2001), pp. 153–65.

———, *'Zieht hin und erforscht das Land'. Die deutsche Palästinaforschung im 19. Jahrhundert* (Göttingen, 2003).

Goren, Haim and Bruno Schelhaas, 'August Petermann vergleicht Flussgefälle', in P. Weigel et al. (eds), *Das Heilige Land in Gotha. Der Verlag Justus Perthes und die Palästinakartographie im 19. Jahrhundert* (Gotha, 2014), pp. 52–3.

———, 'Berghaus's part of Arabia Petraea and Palestine map (1839) and the Royal Geographical Society. Misuse or misunderstanding?', *Terrae Incognitae* 47/2 (2015), pp. 127–41.

———, 'Colonial, religious and scientific mapping. The cartographic practice of Charles William Meredith van de Velde', in J. García-Álvarez and J. Carlos Garcia (eds), *History of Geography and Colonialism* (Lisbon, 2015), pp. 207–23.

———, 'Conrad Schick plant Straßen und Eisenbahnen', in P. Weigel et al. (eds), *Das Heilige Land in Gotha. Der Verlag Justus Perthes und die Palästinakartographie im 19. Jahrhundert* (Gotha, 2014), pp. 116–17.

———, 'On an unknown measuring of Jerusalem in 1823', *Die Erde* 146/1 (2015), pp. 63–78.

Goren, Haim and Petra Weigel, 'Heinrich Lange kartiert Carl Ritters Forschungen zu Jerusalem', P. Weigel et al. (eds), *Das Heilige Land in Gotha. Der Verlag Justus Perthes und die Palästinakartographie im 19. Jahrhundert* (Gotha, 2014), pp. 102–3.

Grabill, Joseph L., *Protestant Diplomacy and the Near East. Missionary Influence on American Policy, 1810–1927* (Minneapolis, 1971).

Grimm, Julius Ludwig, *Palaestina*, 1 : 900.000 (Berlin, 1830).

Haack, Hermann, 'August Petermann (1822–1878)', in H. Haack (ed.), *Schriften zur Kartographie* (Gotha and Leipzig, 1972), pp. 182–90.

Hagedorn, Jürgen, 'August Heinrich Petermann. Der erste große Wissenschaftsmanager der Geographie', *Geographie und Schule* 33/193 (2011), pp. 33–44.

Halbwachs, Maurice, *La Topographie légendaire des Évangiles en Terre Sainte. Étude de mémoire collective* (Paris, 1941).

Hallote, Rachel, *Bible, Map, and Spade. The American Palestine Exploration Society, Frederick Jones Bliss, and the Forgotten Story of Early American Biblical Archaeology* (Piscataway, NJ, 2006).

Handy, Robert T. (ed.), *The Holy Land in American Protestant Life 1800–1948. A Documentary History* (New York, 1981).

Head, Geoffrey, *The Life and Times of John Relly Beard (1800–1876)* (Bolton, 1997).

Heim, Heinrich Jakob, *Dr. Titus Tobler der Palästinafahrer. Ein appenzellisches Lebensbild nach handschriftlichen Quellen* (Zürich and Trogen, 1879).

Helmuth, Carl, *Karte von Palästina nach Robinson, Smith und v. Schubert bearbeitet und in Stein gestochen von Carl Helmuth*, 1 : 600.000 (Halle, 1843).

———, *Plan von Jerusalem nach Robinson und Eli Smith entworfen und lithographiert von C. Helmuth* (Halle, 1843).

Hemprich, Wilhelm Friedrich and Christian Gottfried Ehrenberg, *Naturgeschichtliche Reisen durch Nord-Afrika und West-Asien in den Jahren 1820 bis 1825* (Berlin, Posen, Bromberg, 1828).

Henze, Dietmar, *Enzyklopädie der Entdecker und Erforscher der Erde* (5 vols, Graz 1978–2004, repr. Darmstadt, 2011).

Herbert, Francis, 'The Royal Geographical Society's membership, the map trade, and geographical publishing in Britain 1830 to *ca* 1930. An introductory essay with listing of some 250 fellows in related professions', *Imago Mundi* 35 (1983), pp. 67–95.

Herre, Friedrich, 'Ein Preusse in der Türkei. Helmuth von Moltkes Mission im Osmanischen Reich', *Damals* 19/2 (1987), pp. 131–55.

Hinderlig, Robert and Stefan Sonderegger, 'Titus Tobler (1806–1877). Ein appenzellisches Universalgenie. Neue Ergebnisse aus Nachlassfunden', *Appenzellische Jahrbücher* 121 (1993), pp. 4–94.

Hoffmann, Matthias, 'August Heinrich Petermann (1822–1878). Ein Gothaer Geograph und Kartograph', *International Yearbook of Cartography* XXIX (1989), pp. 85–9.

Hooker, Edward W., *Memoir of Mrs. Sarah L. Huntington Smith, Late of the American Mission in Syria* (New York, 1846).

Hughes, William, *The Illuminated Atlas of Scripture Geography* (London, 1840).

Humboldt, Alexander von, *Central-Asien. Untersuchungen über die Gebirgsketten und die vergleichende Klimatologie* (2 vols, Berlin, 1844).

———, *Cosmos. A Sketch of a Physical Description of the Universe.* Vol. I (trans. E.C. Otté, London, 1848).

———, 'Rapport sur le voyage fait par MM. Ehrenberg et Hemprich en Égypte, Dongola, Syrie, Arabie, et à la pente orientale du plateau de l'Abyssinie, de 1820 à 1824', *Nouvelles Annales des Voyages et des Sciences Géographiques* 36 (1827), pp. 369–97.

Jacotin, Pierre, *Carte topographique de l'Égypte et de plusieurs parties des pays limitrophes, levée pendant l'expédition de l'armée française* (47 sheets, Paris, 1810).

Jochmus, August von, *Der syrische Krieg und der Verfall des Osmanen-Reiches seit 1840. Aktenmässig dargestellt in officiellen, geheimen und vertraulichen Berichten und Urkunden …* (Frankfurt am Main, 1856).

Johnston, Alexander Keith, *The National Atlas of Historical, Commercial, and Political Geography. Accompanied by Maps and Illustrations of the Physical Geography of the Globe by Dr. Heinrich Berghaus and an Ethnological Map of Europe by Dr. Gustaf Kombst* (Edinburgh, 1843).

———, *The Physical Atlas. A Series of Maps and Notes Illustrating the Geographical Distribution of Natural Phenomena* (Edinburgh and London, 1848).

Johnston Limited, W. & A.K. (ed.), *One Hundred Years of Map Making. The Story of W. & A.K. Johnston* (Edinburgh, 1925).

Jones, Yolande, 'British military surveys of Palestine and Syria 1840–1841', *The Cartographic Journal* 10/1 (1973), pp. 29–41.

Karmon, Yehuda, 'An analysis of Jacotin's Map of Palestine', *Israel Exploration Journal* 10/3 (1960), pp. 155–73, 244–53.

Kassner, Carl, 'C. H. W. Mahlmann zum 125. Geburtstage', *Zeitschrift für angewandte Meteorologie* 53 (1936), pp. 273–7.

Kawerau, Peter, *Amerika und die orientalischen Kirchen: Ursprung und Anfang der amerikanischen Mission unter den Nationalkirchen Westasiens* (Berlin, 1958).

Kiepert, Heinrich, *Karte von Palästina. Vorzüglich nach den Itinerarien und Messungen von E. Robinson u. E. Smith. Konstruirt u. gezeichnet von H. Kiepert. 2 Blatt mit 3 Beikarten* (Halle, 1840, published 1841).

———, *Atlas zu Robinson's Palästina* (Halle, 1841).

———, *Karte von Palaestina. Vorzüglich nach den Itinerarien von E. Robinson u. E. Smith, 1838, mit Benutzung den Jacotin'schen Aufnahme, und den Itinerarien von Burckhardt, Buckingham, Irby & Mangles, von Prokesch, Berggren, E. Smith (1834), Bertou, u.a. so wie der Seetzen'schen Karte* (Berlin, 1840).

———, 'Memoir on the maps accompanying this work', in Robinson and Smith, *Biblical Researches* (London), III, First Appendix, pp. 29–55.

———, 'Note über die Construction der Karte zu Consul Wetzstein's Reise', *Zeitschrift für allgemeine Erdkunde* N.F. 7 (1859), pp. 204–8.

———, 'Topographische Beobachtung und Zeichnung', in G. von Neumayer (ed.), *Anleitung zu wissenschaftlichen Beobachtungen auf Reisen* (Berlin, 1875), pp. 39–48.

———, *Übersichts-Karte von Palästina und Phönicien, vorzüglich nach den Messungen und Beobachtungen von Ed. Robinson und Eli Smith, und mit Benutzung aller bis jetzt erschienen Reisewerke neu construirt von H. Kiepert zur Erläuterung von Prof. Ed. Robinson's zweiter Reise in Palästina* (Berlin, 1857).

Kirchberger, Ulrike, *Aspekte deutsch-britischer Expansion. Die Überseeinteressen der deutschen Migranten in Großbritannien in der Mitte des 19. Jahrhunderts* (Stuttgart, 1999).

Kirchhoff, Markus, *Text zu Land. Palästina im wissenschaftlichen Diskurs 1865–1920* (Göttingen, 2005).

Kitto, John, *A Cyclopaedia of Biblical Literature* (2 vols, Edinburgh, 1845).

Klaassen, Marinus J.C., *Adelborstenopleiding te Delft-Medemblik-Breda, 1816–1857* (Den Helder, 1979).

Kombst, Gustaf, *Erinnerungen aus meinem Leben* (Leipzig, 1848).

Krämer, Gudrun, *A History of Palestine: From the Ottoman Conquest to the Founding of the State of Israel* (Princeton, 2011).

Kruse, Friedrich (ed.), *Ulrich Jasper Seetzen's Reisen durch Syrien, Palästina, Phönicien, die Transjordan-Länder, Arabia Petraea und Unter-Aegypten* (4 vols, Berlin, 1854).

Kutluoğlu, Muhammed H., *The Egyptian Question (1831–1841). The Expansionist Policy of Mehmed Ali Paşa in Syria and Asia Minor and the Reaction of the Sublime Porte* (Istanbul, 1998).

Lange, Henry H., *Skizze eines Grundrisses von Jerusalem. Zu C. Ritter's Allg. Erdkunde. Theil XVI, Palaestina, S. 297–508* (Berlin, 1852).

Laughton, J.K., rev. R.O. Morris, 'Washington, John (1800–1863)', ODNB, 28807

Leavy, Margaret R., 'Looking for the Armenians. Eli Smith's missionary adventure, 1830–1831', *Transactions of the Connecticut Academy of Arts and Science* 50 (1992), pp. 189–275.

Lehmann, Edgar, 'Carl Ritters kartographische Leistung', *Die Erde* 90 (1959), pp. 184–222.

Lenz, Karl, 'Heinrich Kiepert. Seine Stellung und Bedeutung in der Geographie', in L. Zögner (ed.), *Antike Welten. Neue Regionen. Heinrich Kiepert 1818–1899* (Berlin, 1999), pp. 11–22.

Lepsius, Karl Richard, *Denkmäler aus Ägypten und Äthiopien* (12 vols, Berlin, 1849–59).

Lindsay, Alexander William Crawford, Lord, *Letters on Egypt, Edom, and the Holy Land*, I–II (London, 1858).

Livingstone, David N., *The Geographical Tradition. Episodes in the History of a Contested Enterprise* (Oxford, Cambridge, 1992).

Lynch, William Francis, *Narrative of the United State's Expedition to the River Jordan and the Dead Sea* (Philadelphia, PA, 1849).

———, *Sketch Map of the River Jordan, from the Observations of the U.S. Expedition under the command of Lt. W. F. Lynch, U.S.N. Narrative of the United States' Expedition to the River Jordan and the Dead Sea* (London, 1848).

———, *Official Report of the United States' Expedition to Explore the Dead Sea and the River Jordan* (Baltimore, 1852).

McCarthy, James, *Journey into Africa. The Life and Death of Keith Johnston, Scottish Cartographer and Explorer (1844–79)* (Latheronwheele, 2004).

McLachlan, Herbert, *Records of a Family 1800–1933. Pioneers in Education, Social Service and Liberal Religion* (Manchester, 1935).

Mahlmann, Wilhelm, 'Bericht über die Witterungsbeobachtungen der Herrn Gen.-Consul v. Wildenbruch, Dr. van Dyck, Dr. de Forest und Hurter in Palästina, nach den von Herrn v. Wildenbruch eingesandten Beobachtungsjournalen', *MVGEB* N.F. 1 (1844), pp. 236–48.

———, 'Russeggers Bemerkungen über dessen meteorologische Hauptdurchschnitte und die barometrische Höhenbestimmung des Todten Meeres', *MVGEB* N.F. 3 (1846), pp. 163–7.

Malte-Brun, Victor Adolphe, 'Map of the Holy Land (review)', *Nouvelles annales des voyages* 18/6 (1859), pp. 353–65.

Makdisi, Ussama, *Artillery of Heaven. American Missionaries and the Failed Conversion of the Middle East* (Ithaca, NY, and London, 2008).

Mansell, Arthur L., *Mediterranean. Coast of Syria, Sheet 3. Ras-en-Nakúra to El Arish.* (London, 1864–5).

Mansell, Arthur L. and Thomas A. Hull (1864). *Mediterranean. Coast of Syria, Sheet 2: Markhab to Ras-en-Nakúra.* (London, 1864).

Marchand, Suzanne L., *German Orientalism in the Age of Empire: Religion, Race, and Scholarship* (Cambridge, 2009).

Marshall-Cornwall, James, 'Three Soldier-Geographers', *Geographical Journal* 131/3 (1965), pp. 357–65.

Meiner, Annemarie, *Geschichte des Verlages Johann Ambrosius Barth Leipzig 1780–1930* (Leipzig, 1930).

Milner, Thomas and August Petermann, *The Atlas of Physical Geography* (London, 1850).

———, *The Atlas of Political Geography* (London, 1851).

———, *A Descriptive Atlas of Astronomy, and of Physical and Political Geography* (London, 1850).

———, *The Library Atlas of Physical and Political Geography* (London, 1854).

Molhuysen, Philippus Christianus and Pieter Johannes Blok (eds), *Nieuw nederlandsch biografisch woordenboek,* 10 vols (Leiden, 1911–37).

Molyneux, Thomas Howard, 'Expedition to the Jordan and the Dead Sea', *JRGS* 18 (1848), pp. 104–30.

Moscrop, John James, *Measuring Jerusalem. The Palestine Exploration Fund and British Interests in the Holy Land* (New York, 2000).

Murphy, David Thomas, *German Exploration of the Polar World. A History, 1870–1940* (Lincoln, 2002).

Noack, Lilian and Dieter Noack, 'Cost of living in Daumier's time'. Compiled by L. & D. Noack since 1999. Last updated: 23 July 2014. Available at http://www.daumier.org/176.0.html (accessed 25 June 2015).

Ogrissek, Rudolf, 'August Petermann (1822–1878). Aus seinem Leben und Wirken als Kartograph', *Vermessungstechnik* 26/11 (1978), pp. 381–3.

Officer, Charles B. and Jake Page, *A Fabulous Kingdom. The Exploration of the Arctic* (Oxford, 2001).

Oren, Michael B., *Power, Faith and Fantasy. America in the Middle East 1776 to the Present* (New York and London, 2007).

Partsch, Joseph, 'Heinrich Kiepert. Ein Bild seines Lebens und seiner Arbeit', *Geographische Zeitschrift* 7 (1901), pp. 1–21, 77–94.

Perthes, Joachim, 'Fünf Generationen', in *Justus Perthes 1785–1935, Hauptkatalog 1935* (Gotha, 1935), pp. VII–XXXII.

Perthus, Sophie and Jutta Faehndrich, 'Visualizing the map-making process. Studying 19th century Holy Land cartography with MapAnalyst', *e-Perimetron* 8/2 (2013), pp. 60–84.

Petermann, August. *Chaînes de montagnes et volcans de l'Asie central selon les observations astronomiques et les mesures hypsomètriques les plus récentes par A. de Humboldt ... terminé par C. Petermann à Potsdam 1841* (Paris, 1843).

———, *Die Erzeugnisse des deutschen Gewerbefleisses in ihrer National-Ausstellung zu Berlin 1844. Nach geographischer Verbreitung der betheiligten Fabrikationsstätten* (Potsdam, 1844).

———, 'Karte der Bai von Súes. Nach der Engl. Aufnahme v. Comm. Mansell, 1856', *PGM* 7 (1861), Tafel 15.

———, 'Karte des Nil-Delta und des Isthmus von Sués von A. Petermann', *PGM* 10 (1864), Tafel 8.

———, 'Note, containing tabular summary, &c.', *JRGS* 20 (1850), pp. 232–5.

———, 'On the fall of the Jordan, and of principal rivers in the United Kingdom', *JRGS* 18 (1848), pp. 89–104.

———, 'Die projektirte Kanalisirung des Isthmus von Sues, nebst Andeutungen über die Höhen-Verhältnisse der angrenzenden Regionen, besonders Palästina', *PGM* 1 (1855), pp. 364–75.

———, 'Vollendung der Jubel-Ausgabe von Stieler's Hand-Atlas in 84 Blättern. Notizen zur Kartographie Palästina's und Polynesiens', *PGM* 14 (1868), pp. 372–7.

Petermann, Heinrich, *Reisen im Orient* (2 vols, Leipzig, 1861).

Phillips, Clifton Jackson, *Protestant America and the Pagan World. The First Half Century of the American Board of Commissioners for Foreign Missions, 1810–1860* (Cambridge, MA, 1969).

Plewe, Ernst (ed.), *Die Carl Ritter Bibliothek* (Wiesbaden, 1978).

Porter, Josias Leslie, *Five Years in Damascus* (2 vols, London, 1855).

———, *The Giant Cities of Bashan and Syria's Holy Places* (London, 1866).

———, *A Handbook for Travellers in Syria and Palestine. Including an Account of the Geography, History, Antiquities, and Inhabitants of these Countries ...* (London, 1858).

———, *Map of Damascus, Hauran and the Lebanon Mountains* (London, 1855).

———, 'Memoir on the map of Damascus, Hauran, and the Lebanon mountains', *JRGS* 26 (1856), pp. 43–55.

Raumer, Karl von, *Beiträge zur biblischen Geographie. Nebst einem Höhendurchschnitt* (Leipzig, 1843).

———, 'Das östliche Palästina und das Land Edom. Zweiter Beitrag zur biblischen Erdkunde', *Annalen der Erd-, Völker- und Staatenkunde* 1 (1829–30), pp. 553–66.

———, *Palästina* (Leipzig, 1838).

———, *Der Zug der Israeliten aus Aegypten nach Canaan. Ein Versuch* (Leipzig, 1837).

Reenders, Hommo, *Alternatieve zending: Ottho Gerhard Heldring (1804–1876) en de verbreiding van het christendom in Nederlands-Indië* (Kampen, 1991).

———, 'Het Réveil en het Heilige Land (1830–1875)', *Jaarboek voor de geschiedenis van het Nederlands protestantisme na 1800* 2 (1994), pp. 50–73.

Religious Tract Society, *The New Biblical Atlas, and Scripture Gazetteer* (London, 1852).

Rey, Emmanuel Guillaume, 'Examen de quelques parties de la carte de la Palestine de M. van de Velde (review)' *Bulletin de la Société de géographie de Paris* 17 (1859), pp. 198–204.

Richter, Hans (ed.), *Carl Ritter. Werk und Wirkungen* (Gotha, 1983).

Richter, Otto, *Der teleologische Zug im Denken Carl Ritters* (Borna and Leipzig, 1905).

Ritchie, George Steve, *The Admiralty Chart. British Naval Hydrography in the Nineteenth Century* (London, 1967).

Ritter, Carl, *The Comparative Geography of Palestine and the Sinaitic Peninsula* (trans. William L. Gage) (4 vols, New York, 1866, repr. 1968).

———, *Die Erdkunde im Verhältnis zur Natur und zur Geschichte des Menschen oder allgemeine vergleichende Geographie als sichere Grundlage des Studiums und Unterrichts in physicalischen und historischen Wissenschaften* (Berlin, 1818–58).

———, *Die Erdkunde von Asien.* Vol. VIII, second unit: *Die Sinai-Halbinsel, Palästina und Syrien.* Second section, first unit: *Palästina und Syrien* (Berlin, 1850).

———, *Der Jordan und die Beschiffung des Todten Meeres* (Berlin, 1850).

———, 'L. v. Wildenbruch's Nivellements in Syrien', *MVGEB* N.F. 3 (1846), pp. 270–2.
———, 'Neue Entdeckungen über die Jordan-Quelle von Major Robe, und über einzelne Punkte von Palästina durch Wolcott, nach Mittheilungen von Herrn Robinson', *MVGEB* 5 (1842–3), pp. 124–7.
———, *Vergleichende Erdkunde der Sinai-Halbinsel, von Palästina und Syrien*, I–IV (Die Erdkunde im Verhältniß zur Natur und zur Geschichte des Menschen, oder allgemeine vergleichende Geographie, XIV–XVII, Berlin, 1848–55).
———, 'Wildenbruch's Bemerkungen über Gaillardot's Karte', *MVGEB* N.F. 3 (1846), pp. 249–52.
Robinson, Arthur H., *Early Thematic Mapping in the History of Cartography* (Chicago and London, 1982).
Robinson, Edward, *Biblical Researches in Palestine and the Adjacent Regions. A Journal of Travels in the Years 1838 & 1852* (3 vols, London, 1856).
———, 'Biographical notices of Gesenius and Nordheimer', *BS* 1/2, pp. 361–90.
———, 'Depression of the Dead Sea and of the Jordan Valley', *JRGS* 18 (1848), pp. 77–88.
———, 'Extracts from a journal of travels in Palestine &c., in 1838; undertaken for the illustration of Biblical Geography', *JRGS* 9 (1839), pp. 295–310.
———, *Physical Geography of the Holy Land* (London, 1865).
———, 'Researches in Palestine. Compiled by the editor from various communications received at different times from the Rev. Eli Smith and the Rev. S. Wolcott. With a Map of the Sources of the Jordan', *BS* 1/1 (1843), pp. 9–88.
———,'Topography of Jerusalem', *BSTR* 3 (1846), pp. 413–61, 605–52.
Robinson, Edward and Eli Smith, *Biblical Researches in Palestine, Mount Sinai and Arabia Petraea. A Journal of Travels in the Year 1838*, I–III (London, 1841).
———, *Biblical Researches in Palestine, Mount Sinai and Arabia Petraea in 1838. A Journal of Travels in the Year 1838*, I–III (Boston, 1841).
———, 'Kurzer Bericht über eine Reise nach Palästina und der Umgegend im Jahr 1838, in Bezug auf die biblische Geographie unternommen', *Zeitschrift für die Kunde des Morgenlandes* 2 (1839), pp. 325–83.
———, *Palästina und die südlich angrenzenden Länder. Tagebuch einer Reise im Jahre 1838 in Bezug auf die biblische Geographie unternommen*, I–III (Halle, 1842).
Robinson, Edward, *Later Biblical Researches in Palestine and the Adjacent Regions. A Journal of Travels in the Year 1852* (London, 1856).
———, *Neuere Biblische Forschungen in Palästina und in den angränzenden Ländern. Tagebuch einer Reise im Jahre 1852* (Berlin, 1857).
Rochfort Scott, Charles, *Map of Syria, Constructed From the Surveys and Sketches of the Undermentioned Officers in that Country in 1840–1841, by Major C. Rochfort Scott, R. Staff-Corps, Under Whose General Direction the Work Was Undertaken, Major F. H. Robe, 87th Fusiliers and R. Wilbraham, 7th Fusiliers, and Lieut. J. F. A. Symonds R[oyal] Engineers* (London, 1846).
Röhricht, Reinhold, *Bibliotheca Geographica Palaestinae. Chronologisches Verzeichnis der auf die Geographie des Heiligen Landes bezüglichen Literatur von 333 bis 1878 und Versuch einer Cartographie* (Berlin, 1890, repr. Jerusalem, 1963, ed. David H.K. Amiran).
Rombach, Johannes H., 'C.W.M. van de Velde (1818–1898), tekenaar reiziger enz.', *De negentiende eeuw. Documentatieblad Werkgroep 19e eeuw* 4/2–3 (1980), pp. 108–21.
———, 'Two great figures in Red Cross history', *International Review of the Red Cross* 16 (1962), pp. 351–61.
Rubin, Rehav (Buni), *Portraying the Land. Hebrew Maps of the Land of Israel from Rashi to the Early 20th Century* (Jerusalem, 2014). [Heb.]

Ruffle, John, 'The Journeys of Lord Prudhoe and Major Orlando Felix in Egypt, Nubia and the Levant 1826–1829', in P. Starkey and J. Starkey (eds), *Travellers in Egypt* (London and New York, 2001), pp. 75–84.

Ruppel, Hans-Rudolf, *Universeller Geist und guter Europäer. Christian Carl Josias von Bunsen 1791–1860. Beiträge zu Leben und Werk des 'Gelehrten Diplomaten' zum 200. Geburtstag* (Korbach, 1991).

Russegger, Joseph von, *Reisen in Europa, Asien und Afrika*. Vol. 3: *Reise in Unter-Egypten, auf der Halbinsel des Sinai und im gelobten Lande* (Stuttgart, 1847).

Ryland, Jonathan Edwards, *Memoirs of John Kitto, Editor of 'the Pictorial Bible' ... Compiled Chiefly from his Letters and Journals, with a Critical Estimate of Dr. Kitto's Life and Writings by Prof. Eadie* (Edinburgh, 1856).

Said, Edward, *Orientalism* (London, 1978).

Sandreczki, Carl, 'Zur alten Geographie Palästina's', *Das Ausland* 5 (1872), pp. 72–8, 97–103.

Saulcy, Louis Felicien de, *Narrative of a Journey Round the Dead Sea and in the Bible Lands in 1850 and 1851. Including an Account of the Discovery of the Sites of Sodom and Gomorrah*, I–II (London, 1854).

Schelhaas, Bruno, 'August Petermann sammelt Höhenmessungen', in P. Weigel et al. (eds), *Das Heilige Land in Gotha. Der Verlag Justus Perthes und die Palästinakartographie im 19. Jahrhundert* (Gotha, 2014), pp. 60–1.

———, 'August Petermann verwertet Kartenentwürfe', in P. Weigel et al. (eds), *Das Heilige Land in Gotha. Der Verlag Justus Perthes und die Palästinakartographie im 19. Jahrhundert* (Gotha, 2014), pp. 54–5.

———, 'August Petermann visualisiert Höhen- und Klimadaten', in P. Weigel et al. (eds), *Das Heilige Land in Gotha. Der Verlag Justus Perthes und die Palästinakartographie im 19. Jahrhundert* (Gotha, 2014), pp. 57–9.

———, '"Das Wiederkehren des Fragezeichens in der Karte". Gothaer Kartenproduktion im 19. Jahrhundert', *Geographische Zeitschrift* 97/1 (2009), pp. 227–42.

Schelhaas, Bruno and Ute Wardenga, '"Inzwischen spricht die Karte für sich selbst". Transformation von Wissen im Prozess der Kartenproduktion', in S. Siegel and P. Weigel (eds), *Die Werkstatt des Kartographen. Materialien und Praktiken visueller Welterzeugung* (München, 2011), pp. 89–108.

Schick, Conrad, 'Studien über Strassen- und Eisenbahn-Anlagen zwischen Jaffa und Jerusalem, nebst Beschreibung der Gegend im Norden von Jaffa und der Ruinen von Mar Zacharias', *PGM* 13 (1867), pp. 124–32.

Schienerl, Jutta, *Der Weg in den Orient. Der Forscher Ulrich Jasper Seetzen. Von Jever in den Jemen (1802–1811)* (Oldenburg, 2000).

Schmidt-Clausen, Kurt, 'Der Beitrag Bunsens zur Gründung des Bistums Jerusalem', in E. Geldbach (ed.), *Der Gelehrte Diplomat. Zum Wirken Christian Carl Josias Bunsens* (Leiden, 1980), pp. 45–63.

Schneider, Karl Friedrich Robert, *Wandkarte zur biblischen Geschichte nach den neuesten Hülfsmitteln, besonders nach den Angaben und Karten von E. Robinson und E. Smith, K. v. Raumer, H. v. Schubert, H. Kiepert [...], bearbeitet. Gezeichnet von E. Braun u. E. Leeder, lithogr. von J. Wilhard in Dresden*, 1 : 500,000 (Dresden, undated).

Schneider, Ulrich Johannes, 'Encyclopedias', in M. Middell (ed.), *Cultural Transfers, Encounters and Connections in the Global 18th Century* (Leipzig, 2014), pp. 163–81.

Schubert, Gotthilf Heinrich von, *Der Erwerb aus einem vergangenen und die Erwartungen von einem zukuenftigen Leben. Eine Selbstbiographie* (5 vols, Erlangen, 1854–6).

———, 'Palästina und die südlich angränzenden Länder. Tagebuch einer Reise im Jahre 1838 ... von E. Robinson und E. Smith ... Halle 1841', *Gelehrte Anzeigen* 13 (1841), pp. 681–6, 689–94, 697–703, 705–11, 713–28, 735–6.

———, *Reise in das Morgenland in den Jahren 1836 und 1837*, I–III (Erlangen, 1839).

Schultz, Ernst Gustav, *Jerusalem. Eine Vorlesung, mit einem Plane gezeichnet von H. Kiepert* (Berlin, 1845).

Schunka, Alexander, 'Das Rohe, das Gekochte – und das Kochrezept. Kartenkommentare des 19. Jahrhunderts als historische Quellen', in S. Siegel and P. Weigel (eds), *Die Werkstatt des Kartographen. Materialien und Praktiken visueller Welterzeugung* (München, 2011), pp. 143–60.

Scully, Richard J., *British Images of Germany. Admiration, Antagonism and Ambivalence, 1860–1914* (Basingstoke, 2012).

———, '"North Sea or German Ocean"? The Anglo-German cartographic freemasonry, 1842–1914', *Imago Mundi* 62/1 (2010), pp. 46–62.

Secord, James A., *Victorian Sensation. The Extraordinary Publication, Reception and Secret Authorship of 'Vestiges of the Natural History of Creation'* (Chicago, 2000).

Seetzen, Ulrich Jasper, *Verzeichniss der für die orientalische Sammlung in Gotha zu Damask, Jerusalem u. s. w. angekauften orientalischen Manuscripte und gedruckten Werke, Kunst- und Naturprodukte u. s. w.*, ed. Ernst Friedrich Karl Rosenmüller (Leipzig, 1810).

Sides, Hampton, *In the Kingdom of Ice. The Grand and Terrible Polar Voyage of the U.S.S.* Jeanette (New York, 2014).

Silberman, Neil Asher, *Digging for God and Country. Exploration, Archaeology, and the Secret Struggle for the Holy Land, 1799–1917* (New York, 1982).

Smith, David, 'The business of W. & A.K. Johnston, 1826–1901', *IMCoS Journal* 82 (2000), pp. 9–19.

Smith, Eli, 'An ancient temple on Mt. Lebanon', *BS* 1 (1843), pp. 557–63.

———, 'The Wines of Mount Lebanon', *BSTR* 3 (1846), pp. 385–9.

Smith, Eli and Harrison Gray Otis Dwight, *Missionary Researches in Armenia Including a Journey through Asia Minor, and into Georgia and Persia, with a Visit to the Nestorian and Chaldean Christians of Oormiah and Salmas* (London, 1834).

Smith, George Adam, *The Historical Geography of the Holy Land* (London, 1974 [1894]).

———, *The Life of John Wilson. D.D. F.R.S. For Fifty Years Philanthropist and Scholar in the East* (London, 1878).

Smith, Henry Boynton and Roswell Dwight Hitchcock, *The Life, Writings and Character of Edward Robinson* (New York, 1977 [1863]).

Smits, Jan, *Petermann's Maps. Carto-Bibliography of the Maps in Petermanns Geographische Mitteilungen 1855–1945* ('t Goy-Houten, 2004).

Society for Diffusion of Useful Knowledge, *Maps of the Society for the Diffusion of Useful Knowledge*, Vol. 1 (London, 1844).

Soler, Renaud, *Edward Robinson (1794–1863) et l'émergence de l'archéologie biblique* (Paris, 2014).

Stanley, Arthur Penrhyn, *Sinai and Palestine, in Connection With Their History* (London, 1856). Available at http://www.biblicalcyclopedia.com (accessed 17 August 2016).

Steudner, Hermann 'Die Deutsche Expedition bei den Moses-Quellen im Peträischen Arabien, 26. bis 31. Mai 1861. Nach einem Briefe Dr. Steudner's vom 4. Juni 1861', *PGM* 7 (1861), pp. 427–9.

Stieler, Adolf, *Hand-Atlas über alle Theile der Erde* (Gotha, 1817–1945).

Stresemann, Erwin, *Hemprich und Ehrenberg. Reisen zweier naturforschender Freunde im Orient, geschildert in ihren Briefen aus den Jahren 1819–1826* (Berlin, 1954).

Strong, James and McClintock, John, *The Cyclopedia of Biblical, Theological, and Ecclesiastical Literature* (New York, 1880).

Thomson, William M., *The Land and the Book or Biblical Illustrations Drawn From the Manners and Customs, the Scenes and the Scenery of the Holy Land.* 2 vols (New York, 1859).

Tibawi, Abdul Latif, *American Interests in Syria 1800–1901: A Study of Educational, Literary and Religious Work* (Oxford, 1966).

Tobler, Titus, *Bibliographia geographica Palaestina. Zunächst kritische Übersicht gedruckter und ungedruckter Beschreibungen der Reisen ins heilige Land* (Leipzig, 1867).
———, *Planography of Jerusalem. Memoir to Accompany the New Ground-Plan of the City of Jerusalem …* (Gotha, 1858).
———, *Titus Toblers dritte Wanderung nach Palästina im Jahre 1857. Ritt durch Philistäa, Fußreisen im Gebirge Judäas und Nachlese in Jerusalem* (Gotha, 1859).
Tracy, Joseph, *Palestine and Mount Lebanon. Enlarged from Kiepert's Map, in Robinson and Smith's Biblical Researches, with Additions and Corrections, Principally from the Same Authorities* (Boston, [1845]).
Tresse, René, 'L'installation du premier consul d'Angleterre à Damas (1830–1834)', *Revue d'histoire des colonies* 24 (1936), pp. 359–80.
Trumbull, H. Clay, *Kadesh-Barnea, its Importance and Probable Site with the Story of a Hunt for it Including Studies of the Route of the Exodus and the Southern Boundary of the Holy Land* (New York, 1884).
Union Theological Seminary [Board of Directors], *The Union Theological Seminary, in the City of New York, its History, Condition and Wants: An Appeal to the Friends of the Institution* (New York, 1852).
van de Velde, Charles William Meredith, *Carte de la Terre Sainte pour servir à l'étude des Saintes Ecritures* (Paris, 1865).
———, 'C. W. M. van de Velde's letzte Reise in Palästina, 1861/62, und Bericht über die neue Auflage seiner Karte des Heiligen Landes', *PGM* 11 (1865), pp. 188–91, 215–24, 296–305.
———, *Kaart van het Eiland Java te zamengesteld uit officiele bronnen door door C.W.M. van de Velde, Luitenant ter Zee* (Leiden, 1845).
———, *Karte von Palästina. Deutsche Ausgabe nach der 2. Auflage der 'Map of the Holy Land'* (Gotha, 1866).
———, *Map of the Holy Land Constructed From His Own Surveys in 1851, 52 & 62. From Those Made in 1841 by Majors Robe, Rochfort Scott, & Lieut. Symonds … From the Triangulations Made in 1860–62 Under Direction of Commr. A.L. Mansell, R.N.; and From the Researches of Robinson, Wetzstein and Other Travellers. Engraved by Eberhardt and Stichardt* (Gotha, 1865).
———, *Map of the Holy Land. From his own surveys in 1851 & 1852, from those made in 1841 by Majors Robe and Rochfort Scott, Lieut. Symonds … and from the Results of the Researches made by Lynch, Robinson, Wilson, Burckhardt, Seetzen & c. Engraved by Eberhardt and by Stichardt* (Gotha, 1858).
———, *Memoir to Accompany the Map of the Holy Land* (Gotha, 1858).
———, *Narrative of a Journey through Syria and Palestine in 1851 and 1852*, 2 vols (Edinburgh and London, 1854).
———, *Le pays d'Israël. Collection de cent vues prises d'après nature dans la Syrie et la Palestine* (Paris, 1857).
———, *Plan of the Town and Environs of Jerusalem Constructed from the English Ordnance-Survey and Measurements of Dr. Titus Tobler, with Memoir by Dr. Titus Tobler* (Gotha and London, 1858).
———, *Reis door Syrië en Palestina in 1851 en 1852* (Utrecht, 1854).
———, *Reise durch Syrien und Palästina in den Jahren 1851 und 1852* (Leipzig, 1855–6).
———, 'Van de Velde's Karte des Heiligen Landes', *PGM* 11 (1865), Tafel 7.
———, *Vues de Java, Sumatra, Borneo, Celebes et autres possessions hollandaises dans les Grandes Indes, dessinées d'après nature* (Amsterdam, 1846).
van Lieburg, Fred, 'Een wereldwijde opwekking in het protestant vaderland. De historiografie van het Nederlands Réveil', in F. van Lieburg (ed.), *Opwekking van de natie: Het protestantse Réveil in Nederland* (Hilversum, 2012). pp. 241–80.
de Vries, Dirk, 'De kaart van Palestina door C.W.M. van de Velde (1858): een 'autoriteitsstuk', in P. van der Krogt (ed.), *'Capita selecta' uit de geschiedenis van de kartografie* (Amersfoort, 1996), pp. 15–26.

Wagstaff, Malcolm, 'Colonel Leake. Traveler and scholar', in S. Searight and M. Wagstaff (eds), *Travellers in the Levant* (Durham, 2001), pp. 3–15.

———, 'Leake, William Martin', *ODNB*, 16242.

Wallach, Jehuda L., *Anatomie einer Militärhilfe. Die preußisch-deutschen Militärmission in der Türkei 1835–1919* (Düsseldorf, 1976).

Washington, John, 'Admiralty surveys', *JRGS* 31 (1861), pp. cxxxiv–cxlvi.

Webster, Diane, 'Maps', in B. Bell., D. Finkelstein and A. McCleery (eds), *The Edinburgh History of the Book in Scotland.* Vol. 3: *Ambition and Industry, 1800–1880* (Oxford, 2007), pp. 313–23.

Weigel, Petra, 'August Petermann modernisiert alte Kartenbilder', in P. Weigel et al. (eds), *Das Heilige Land in Gotha. Der Verlag Justus Perthes und die Palästinakartographie im 19. Jahrhundert* (Gotha, 2014), pp. 62–5.

———, 'Heinrich Berghaus verknüpft Zahlen, Karten und Berichte', in P. Weigel et al. (eds), *Das Heilige Land in Gotha. Der Verlag Justus Perthes und die Palästinakartographie im 19. Jahrhundert* (Gotha, 2014), pp. 44–5.

———, 'Karl Georg von Raumer betreibt Bibelkartographie', in P. Weigel et al. (eds), *Das Heilige Land in Gotha. Der Verlag Justus Perthes und die Palästinakartographie im 19. Jahrhundert* (Gotha, 2014), pp. 50–1.

———, *Die Sammlung Perthes Gotha* (Berlin, 2011).

Weigel, Petra, Haim Goren, Bruno Schelhaus and Jutta Faehndrich (eds), *Das Heilige Land in Gotha. Der Verlag Justus Perthes und die Palästinakartographie im 19. Jahrhundert* (Gotha, 2014).

Weigel Williams, Hermine (ed.), *Robinson's Letter-Journal (1826–1829). Written from Europe by Edward Robinson to His Sister Elisabeth* (New York and Bloomington, 2009).

Weller, Ewald, *August Petermann. Ein Beitrag zur Geschichte der geographischen Entdeckungen und der Kartographie* (Leipzig, 1911).

Westphal, Johann Heinrich, 'Ueber die topographische Lage Jerusalems, zur Erläuterung des Planes aufgenommen von dem Herrn Dr. Westphal in Göttingen. Aus dessen Tagebuche während einer Reise durch den Orient in den Jahren 1822 und 1823', *Hertha* 1/3 (1825), pp. 385–90.

Wichmann, Hugo, 'Petermann, August', *Allgemeine Deutsche Biographie* 26 (1888), pp. 795–805.

Wildenbruch, Louis von, 'Notes on the Physical Geography of Palestine', *JRGS* 20 (1850), pp. 227–30.

Williams, George, *The Holy City. Historical, Topographical, and Antiquarian Notices on Jerusalem*, I–II (London, 1849).

Williams, Jay Gomer, *The Education of Edward Robinson* (New York, 1997).

———, *The Times and Life of Edward Robinson. Connecticut Yankee in King Solomon's Court* (Atlanta, 1999).

Wilson, Charles W., 'Recent surveys in Sinai and Palestine', *JRGS* 43 (1873), pp. 206–40.

Wilson, John, *The Lands of the Bible, Visited and Described in an Extensive Journey Undertaken with Special Reference to the Promotion of Biblical Research and the Advancement of the Cause of Philanthropy* (2 vols, Edinburgh and London, 1847).

———, 'Map of the Holy Land with topographical plans of Jerusalem, Tyre, Sidon, Akka, and Beirut, constructed for Dr Wilson's "Lands of the Bible". By W. & A. K. Johnston, geographers and engravers to the Queen', in J. Wilson, *The Lands of the Bible, Visited and Described in an Extensive Journey Undertaken with Special Reference to the Promotion of Biblical Research and the Advancement of the Cause of Philanthropy.* Vol. II (Edinburgh and London, 1847), facing p. 416.

———, 'Map of the Peninsula of Mount Sinai and Arabia Petraea, constructed for Dr Wilson's "Lands of the Bible". By W. & A. K. Johnston, Geographers and Engravers to the Queen [side maps: Valley and Rocks of Petra from Laborde; Central Group of the Mountains of Sinai]', in J. Wilson, *The Lands of the Bible, Visited and Described in an Extensive Journey Undertaken with Special Reference to the Promotion of Biblical Research and the Advancement of the Cause of Philanthropy*. Vol. I (Edinburgh and London, 1847), facing p. 160.

Winer, Georg Benedict, *Biblisches Realwörterbuch zum Handgebrauch für Studirende, Kandidaten, Gymnasiallehrer und Prediger* (Leipzig, 1820).

Wolcott, Samuel, 'Geographical notes on Palestine. The coast of the Dead Sea, Masada, route from Mount Lebanon to Baalbek', *BS* 3 (1846), pp. 398–402.

———, 'Maps of Palestine', *BSTR* 2/7 (August 1845), pp. 585–9.

———, 'The topography of Jerusalem', *BS* 23 (1866), pp. 684–96.

Zimmermann, Carl, *Atlas von Palaestina und der Sinai-Halbinsel zu C. Ritters Erdkunde Band XIV–XVI* (Berlin, 1850).

Zimpel, Charles Franz, *Die Israeliten in Jerusalem. Eine Denkschrift gewidmet allen Israeliten* (Stuttgart, 1852).

———, *Neue örtliche topographische Beleuchtung der heiligen Weltstadt Jerusalem, mit besonderer Rücksicht auf die Leidenstage unseres Herrn Jesu Christi und die Identität der heiligen Orte daselbst ...* (Stuttgart, 1853).

———, *Straßen-Verbindung des Mittelländischen mit dem Todten Meere und Damascus über Jerusalem ...* (Frankfurt am Main, 1865).

Zögner, Lothar (ed.), *Antike Welten. Neue Regionen. Heinrich Kiepert 1818–1899* (Berlin, 1999).

——— (ed.), *Carl Ritter in seiner Zeit 1779–1859* (Berlin, 1979).

Index

www.ingramcontent.com/pod-product-compliance
Ingram Content Group UK Ltd.
Pitfield, Milton Keynes, MK11 3LW, UK
UKHW051450290926
15471UKWH00008B/1121

9 781784 534547